Spring Boot 2
企业应用实战

疯狂软件 编著

电子工业出版社
Publishing House of Electronics Industry
北京·BEIJING

内 容 简 介

本书介绍了 Java EE 领域的全新开源框架：Spring Boot 2。本书的示例建议在 Tomcat 8 上运行。

本书重点介绍如何使用 Spring Boot 进行 Java EE 快速开发，从内容上可以划分为四个部分。第一部分详细介绍了 Spring Boot 的核心知识。第二部分详细介绍了 Spring Boot 的 Web 开发。第三部分重点介绍了 Spring Boot 的数据访问。第四部分重点介绍了 Spring Boot 的 Spring Security 安全控制。书中示范开发了一个包含 7 个表、表之间具有复杂的关联映射关系，且业务功能也相对完善的信息管理系统案例，希望让读者理论联系实际，将 Spring Boot 框架真正运用到实际开发当中去。该案例采用目前最流行、最规范的 Java EE 架构，整个应用分为 DAO 持久层、领域对象层、业务逻辑层、控制器层和视图层，各层之间分层清晰，层与层之间以松耦合的方法组织在一起。所有代码完全基于 Eclipse IDE 来完成，一步步带领读者深入两个框架的核心。

阅读本书之前，建议先阅读疯狂软件教育的《疯狂 Java 讲义》一书。本书适合有较好的 Java 编程基础，JSP、Servlet、JDBC 基础，Spring 框架基础的读者，尤其适合于对 Spring Boot 了解不够深入，或对 Spring Boot 整合开发不太熟悉的开发人员阅读。

未经许可，不得以任何方式复制或抄袭本书之部分或全部内容。
版权所有，侵权必究。

图书在版编目（CIP）数据

Spring Boot 2 企业应用实战 / 疯狂软件编著. —北京：电子工业出版社，2018.6
ISBN 978-7-121-34116-8

Ⅰ．①S… Ⅱ．①疯… Ⅲ．①JAVA 语言—程序设计 Ⅳ．①TP312.8

中国版本图书馆 CIP 数据核字（2018）第 083111 号

策划编辑：张月萍
责任编辑：刘 舫
印　　刷：北京盛通商印快线网络科技有限公司
装　　订：北京盛通商印快线网络科技有限公司
出版发行：电子工业出版社
　　　　　北京市海淀区万寿路 173 信箱　　　　邮编：100036
开　　本：787×1092　1/16　　印张：16　　字数：410 千字
版　　次：2018 年 6 月第 1 版
印　　次：2021 年 1 月第 7 次印刷
定　　价：58.00 元

凡所购买电子工业出版社图书有缺损问题，请向购买书店调换。若书店售缺，请与本社发行部联系，联系及邮购电话：(010) 88254888，88258888。

质量投诉请发邮件至 zlts@phei.com.cn，盗版侵权举报请发邮件至 dbqq@phei.com.cn。
本书咨询联系方式：010-51260888-819，faq@phei.com.cn。

前　言

　　时至今日，以 Spring 为核心的轻量级 Java EE 企业开发平台在企业开发中占有绝对的优势，Java EE 应用以其稳定的性能、良好的开放性以及严格的安全性，深受企业应用开发者的青睐，应用的性能、稳定性都有很好的保证。

　　Spring 在 Java EE 开发中是实际意义上的标准，但是在实际项目开发中使用 Spring 的时候经常遇到两个让人非常头疼的问题：

（1）大量的配置文件

（2）与第三方框架整合

　　特别是在今天，脚本语言和敏捷开发大行其道之时，Spring 的开发显得尤其烦琐。而 Spring Boot 的推出正具有颠覆和划时代的意义。如果说 Spring 框架的目标是帮助开发者写出更好的系统，那 Spring Boot 的目标就是帮助开发者用更少的代码，更快地写出好的系统。

　　Spring Boot 从无数知名企业的实践中吸取经验，其设计目的是用来简化新 Spring 应用的初始搭建以及开发过程。Spring Boot 遵循"约定优于配置"原则，从而使开发人员不再需要定义样板化的配置，只需要很少的配置，或者大部分时候只是使用默认配置就可以快速搭建项目，无须配置整合第三方框架。通过这种方式，Spring Boot 在蓬勃发展的快速应用开发（rapid application development）领域已经成为领导者。

　　本书基于 Spring Boot 2.0 版本，重点介绍 Spring Boot 框架，采用 Tomcat 8 作为 Web 服务器，Eclipse IDE 作为开发工具，详细介绍了 Spring Boot 框架的绝大部分功能。希望读者在阅读、学习完本书之后，能够掌握 Spring Boot 技术，更快更好地开发出 Java EE 项目，为 Java 开发者带来更多的就业机会与竞争力。

本书有什么特点

　　本书是一本介绍 Spring Boot 框架的实用图书，全面介绍了最新的 Spring Boot 和常用第三方框架整合等各方面的知识。

　　本书针对每一个知识点都通过相应的程序给出了示范，第 7 章的实战项目"信息管理系统"采用目前非常流行、规范的 Java EE 架构，整个应用分为 DAO 持久层、领域对象层、业务逻辑层、控制器层和视图层，各层之间分层清晰，层与层之间以松耦合的方法组织在一起。

　　笔者既担任过软件开发的技术经理，也担任过软件公司的培训导师，如今从事专业、高端的职业技术培训，所有应用范例都密切契合企业开发实际场景，例如用户权限验证、文件上传下载等都是企业开发中的实际功能，同时采用了目前企业最流行、最规范的开发架构，严格遵守 Java EE 开发规范。读者参考本书的架构，完全可以身临其境地感受企业实际项目开发。

　　本书并不是一本关于所谓"思想"的书，也没有一堆"深奥"的新名词和"高深"的理念，只会让读者学会实际的 Spring 和 Spring Boot 技术。本书的特点是操作步骤详细，编程思路清晰，语言平实易懂。只要读者认真阅读本书，并掌握书中知识，那么就完全可以胜任企业中的 Spring Boot 项目开发。

阅读本书需要具备一定的计算机知识以及编程功底。熟练掌握Java语言和Spring框架的IoC、AOP和持久层的ORM设计模式等知识对于学习本书是很有必要的。

可访问www.crazyit.org或www.broadview.com.cn/34116下载本书配套资源。

本书写给谁看

如果你已经掌握Java SE的内容，或已经学完疯狂软件教育的《疯狂Java讲义》一书，那么非常适合阅读此书。除此之外，如果你已有初步的JSP、Servlet、JDBC基础，甚至对Spring、Spring Boot等框架有所了解，但希望掌握它们在实际开发中的应用，本书也将非常适合你。如果你对Java的掌握还不熟练，则建议遵从学习规律，循序渐进，暂时不要购买、阅读此书，而是按照"疯狂Java学习路线图"中的建议顺序学习。

衷心感谢

衷心感谢李刚老师，他是一位非常好的朋友，在本书的创作过程中，他提供了大量切实、有用的帮助。同时衷心感谢疯狂软件教育中心所有同事提供的帮助。

感谢所有参加疯狂软件实训的学生，他们在实际工作场景的应用证明了本书的价值，他们的反馈让本书更加实用。

编者
2018年3月1日

目 录 CONTENTS

- 第 1 章 Spring Boot 入门 1
  - 1.1 Spring 简介 2
    - 1.1.1 Spring 概述 2
    - 1.1.2 Spring 的生态圈 3
    - 1.1.3 Spring 5 的变化 4
    - 1.1.4 Spring 的配置简化 4
  - 1.2 Spring Boot 简介 4
    - 1.2.1 Spring Boot 概述 4
    - 1.2.2 Spring Boot 解决的问题 5
    - 1.2.3 Spring Boot 的主要特性 5
    - 1.2.4 Spring Boot 2.0 的重要改变 5
  - 1.3 "开箱即用"的依赖模块 5
    - 1.3.1 日志依赖模块 spring-boot-starter-logging 6
    - 1.3.2 Web 开发依赖模块 spring-boot-starter-web 7
  - 1.4 开发第一个 Spring Boot 应用 8
    - 1.4.1 下载和安装 Maven 8
    - 1.4.2 Eclipse 集成 Maven 9
    - 1.4.3 示例：第一个 Spring Boot 应用 ... 10
  - 1.5 本章小结 18

- 第 2 章 Spring Boot 核心 19
  - 2.1 Spring Boot 的启动类与核心注解 @SpringBootApplication 20
  - 2.2 Spring Boot 基本配置介绍 21
    - 2.2.1 关闭某个自动配置 21
    - 2.2.2 定制启动 banner 22
    - 2.2.3 应用的全局配置文件 23
    - 2.2.4 Spring Boot 的依赖模块 24
  - 2.3 Spring Boot 自动配置原理 25
    - 2.3.1 源码分析 25
    - 2.3.2 spring.factories 分析 27
    - 2.3.3 Spring Boot Web 开发的自动配置 ... 29
  - 2.4 本章小结 30

- 第 3 章 Spring Boot 的 Web 开发 31
  - 3.1 Spring Boot 的 Web 开发支持 32
  - 3.2 Thymeleaf 模板引擎 32
    - 3.2.1 Thymeleaf 概述 33
    - 3.2.2 Thymeleaf 基础语法 33
  - 3.3 Spring 和 Thymeleaf 的整合 36
  - 3.4 Spring Boot 的 Thymeleaf 支持 37
  - 3.5 Spring Boot 的 Web 开发实例 38
    - 示例：第一个 Spring Boot 的 Web 应用 ... 38
    - 示例：Thymeleaf 常用功能 45
  - 3.6 Spring Boot 对 JSP 的支持 52
    - 示例：Spring Boot 添加 JSP 支持 53
  - 3.7 Spring Boot 处理 JSON 数据 57
    - 示例：Spring Boot 处理 JSON 57
  - 3.8 Spring Boot 文件上传下载 63
    - 示例：Spring Boot 文件上传 63
    - 示例：使用对象方式接收上传文件 ... 66
    - 示例：文件下载 69
  - 3.9 Spring Boot 的异常处理 71
    - 示例：ExceptionHandler 处理异常 ... 71
    - 示例：父类 Controller 处理异常 73
    - 示例：Advice 处理异常返回 JSON ... 76
  - 3.10 本章小结 78

- 第 4 章 Spring Boot 的数据访问 79
  - 4.1 Hibernate/JPA/Spring Data JPA 的概念 ... 80
    - 4.1.1 对象/关系数据库映射（ORM） .. 80
    - 4.1.2 基本映射方式 81
    - 4.1.3 流行的 ORM 框架简介 82
  - 4.2 Spring Data JPA 83
    - 4.2.1 Spring Data 核心数据访问接口 ... 83
    - 示例：CrudRepository 接口访问数据 ... 84
    - 示例：PagingAndSortingRepository 接口访问数据 91
    - 4.2.2 Spring Data JPA 开发 99
    - 示例：简单条件查询 100
    - 示例：关联查询和@Query 查询 105
    - 示例：@NamedQuery 查询ata 114
    - 示例：Specification 查询 118
  - 4.3 Spring Boot 使用 JdbcTemplate 128

示例：JdbcTemplate 访问数据 128
4.4 Spring Boot 整合 MyBatis..................... 135
　　示例：Spring Boot 整合 MyBatis 开发..... 135
4.5 本章小结 141

第 5 章　Spring Boot 的热部署与单元测试 142

5.1 使用 spring-boot-devtools 进行热部署..... 143
　　示例：使用 spring-boot-devtools 实现
　　　　　热部署 143
5.2 Spring Boot 的单元测试 147
　　示例：使用 Spring Boot 的单元测试 147
5.3 本章小结 155

第 6 章　Spring Boot 的 Security 安全控制 156

6.1 Spring Security 是什么 157
6.2 Spring Security 入门.............................. 157
　　6.2.1　Security 适配器 157
　　6.2.2　用户认证 158
　　6.2.3　用户授权 158
　　6.2.4　Spring Security 核心类 160
　　6.2.5　Spring Security 的验证机制..... 161
　　6.2.6　Spring Boot 的支持 161
　　示例：简单 Spring Boot Security 应用 ... 162
6.3 企业项目中的 Spring Security 操作......... 173
　　示例：基于 JPA 的 Spring Boot Security
　　　　　操作 173
　　示例：基于 MyBatis 的 Spring Boot
　　　　　Security 操作 180
　　示例：基于 JDBC 的 Spring Boot
　　　　　Security 操作 183
6.4 本章小结 185

第 7 章　实战项目：信息管理系统................. 186

7.1 项目简介及系统架构............................. 187
　　7.1.1　系统功能介绍......................... 187
　　7.1.2　相关技术介绍......................... 187
　　7.1.3　系统结构 188
　　7.1.4　系统的功能模块..................... 188
7.2 配置文件 189
7.3 持久化类 191
　　7.3.1　设计持久化实体..................... 191
　　7.3.2　创建持久化实体类 192
　　7.3.3　导入初始数据......................... 197
7.4 定义 Repository 接口实现 Repository
　　持久层 198
7.5 实现 Service 持久层.............................. 200
　　7.5.1　业务逻辑组件的设计.................. 201
　　7.5.2　实现业务逻辑组件 201
　　7.5.3　事务管理 224
7.6 实现 Web 层.. 224
　　7.6.1　控制器 224
　　7.6.2　系统登录 225
　　7.6.3　菜单管理 233
　　7.6.4　角色管理 235
　　7.6.5　用户管理 240
　　7.6.6　功能扩展 245
7.7 本章小结 249

CHAPTER

1

第 1 章
Spring Boot 入门

本章要点

- Spring 简介
- Spring Boot 简介
- 下载和安装 Maven
- Eclipse 集成 Maven
- Eclipse 构建基于 Maven 的 Spring Boot 项目

时至今日，轻量级 Java EE 平台在企业开发中占有绝对的优势，Java EE 应用以其稳定的性能、良好的开放性以及严格的安全性，深受企业应用开发者的青睐。实际上，对于信息化要求较高的行业，如银行、电信、证券以及电子商务等行业，都不约而同地选择了 Java EE 开发平台。

而 Spring 在 Java EE 开发中已经是实际意义上的标准，绝大部分的软件开发公司在开发 Java EE 应用的时候都会用到 Spring。不仅如此，围绕 Spring，以 Spring 为核心还衍生出了一系列框架，如 Spring Boot、Spring Cloud、Spring Cloud Data Flow、Spring Web Flow、Spring Security、Spring for Android 等（登录 Spring 官方网站 https://spring.io/ 了解具体内容），Spring 越来越强大，在开发中带来越来越多的便捷。本书所介绍的正是现阶段非常流行的 Spring Boot 框架。

1.1 Spring 简介

Spring 框架由 Rod Johnson 开发，是一个轻量级的企业级开发一站式解决方案。2004 年发布了 Spring 框架的第一个版本。经过十多年的发展，Spring 已经发展成 Java EE 开发中最重要的框架之一。对于 Java 开发者来说，Spring 已经成为必须掌握的框架。

▶▶ 1.1.1 Spring 概述

Spring 是一个从实际开发中抽取出来的框架，因此它完成了大量开发中的通用步骤，留给开发者的仅仅是与特定应用相关的部分，从而大大提高了企业应用的开发效率。

Spring 为企业应用的开发提供了一个轻量级的解决方案。该解决方案包括：基于依赖注入的核心机制、基于 AOP 的声明式事务管理、与多种持久层技术的整合以及优秀的 Web MVC 框架等。Spring 致力于 Java EE 应用各层的解决方案，而不是仅仅专注于某一层的方案。可以说，Spring 是企业应用开发的"一站式"选择，Spring 贯穿表现层、业务层、持久层。然而，Spring 并不想取代那些已有的框架，而是以高度的开放性与它们无缝整合。

总结起来，Spring 具有如下优点。

➢ 低侵入式设计，代码的污染极低。

➢ 独立于各种应用服务器，基于 Spring 框架的应用，可以真正实现 Write Once，Run Anywhere 的承诺。

➢ Spring 的 IoC 容器降低了业务对象替换的复杂性，提高了组件之间的解耦。

➢ Spring 的 AOP 支持允许将一些通用任务如安全、事务、日志等进行集中式处理，从而提供了更好的复用性。

➢ Spring 的 ORM 提供了与第三方持久层框架的良好整合，并简化了底层的数据库访问。

➢ Spring 的高度开放性，并不强制应用完全依赖于 Spring，开发者可自由选用 Spring 框架的部分或全部。

图 1.1 显示了 Spring 框架的组成结构图。

正如图 1.1 所示，Spring 是模块化的，这意味着开发者可以选择自己需要的 Spring 模块。其中最重要的模块包括 Core Container（核心容器）、AOP（面向切面编程）、Web（Web 集成功能）、Data Access（数据访问功能）等。

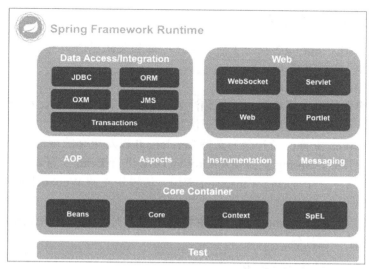

图 1.1　Spring 框架的组成结构图

当使用 Spring 框架时，必须使用 Spring Core Container（即核心容器），它代表了 Spring 框架的核心机制，Spring Core Container 主要由 org.springframework.core、org.springframework.beans、org.springframework.context 和 org.springframework.expression 四个包及其子包组成，主要提供 Spring IoC 容器支持。

▶▶ 1.1.2　Spring 的生态圈

Spring 发展到今天已经不仅仅是 Spring 框架本身的内容了，Spring 目前提供了非常多的基于 Spring 的项目，可以用来更加快捷地开发项目。

Spring 目前主要有以下项目。

- Spring Boot：使用默认 Java 配置来实现快速开发。
- Spring XD：用于简化大数据应用的开发。
- Spring Cloud：为分布式系统开发提供工具集。
- Spring Data：对主流的关系型数据库和 NoSQL 数据库提供支持。
- Spring Integration：通过消息机制对企业集成模式（EIP）提供支持。
- Spring Batch：简化及优化大量数据的批处理操作。
- Spring Security：通过认证和授权保护应用。
- Spring HATEOAS：基于 HATEOAS 原则简化 REST 服务开发。
- Spring Social：与社交网络 API（如 Facebook、新浪微博等）的集成。
- Spring AMQP：对基于 AMQP 的消息的支持。
- Spring Mobile：提供对手机设备进行检测的功能，给不同的设备返回不同的支持。
- Spring for Android：提供在 Android 上消费 RESTful API 的功能。
- Spring Web Flow：基于 Spring MVC 提供的 Web 应用开发。
- Spring Web Services：提供基于协议有限的 SOAP/Web 服务。
- Spring LDAP：简化使用 LDAP 开发。
- Spring Session：提供一个 API 及实现来管理用户会话信息。

1.1.3 Spring 5 的变化

与 Spring 4.x 相比，Spring 5 发生了一些变化，这些变化包括如下这些。
- Spring 5 已经全面支持 Java 8。
- 删除了一些已过时的包和类。
- 核心 IoC 容器新增了泛型限定式依赖注入、Map 依赖注入、List（数组）注入、延迟注入等功能，这些功能主要体现在基于注解的配置上。
- Spring 5 的 Web 支持已经升级为支持 Servlet 3.0 以及更高的规范。
- 从 Spring 5 开始，Spring 支持使用 Groovy DSL 进行 Bean 配置。
- Spring 5 新增了一个 spring-websocket 模块，该模块支持 WebSocket、SockJS、STOMP 通信。

从以上介绍可以看出，Spring 5 的升级主要就是为了支持 Java 8 和 Servlet 3.0 规范，也为核心 IoC 容器增强了一些注解。

1.1.4 Spring 的配置简化

Spring 最大的特点是通过配置简化开发。
- 在 Spring 1.x 的时代，使用 Spring 开发都是使用 xml 配置 Bean。
- 在 Spring 2.x 的时代，Spring 提供了声明 Bean 的注解（如@Component、@Service 等），大大减少了 XML 的配置量。随之而来的是开发者的讨论：注解配置和 XML 配置到底哪个更好？开发者最终的选择是应用的基本配置（如数据源、事务管理）使用 XML，业务配置使用注解。
- 从 Spring 3.x 到今天，Spring 提供了 Java 配置的能力，使用 Java 配置可以让开发者更加理解所配置的 Bean。Spring 5 和 Spring Boot 都推荐使用 Java 的注解配置，所以本书都将使用 Java 的注解配置。

1.2 Spring Boot 简介

1.2.1 Spring Boot 概述

Spring 框架非常优秀，然而它最大的问题在于"配置过多"。基于 Spring 的企业级开发项目，需要大量的配置文件，Spring Boot 的出现就是为了解决 Spring 框架存在的问题。

Spring Boot 是由 Pivotal 团队提供的全新框架，其设计目的是用来简化 Spring 应用的创建、运行、调试、部署等。使用 Spring Boot 可以做到专注于 Spring 应用的开发，而无须过多关注 XML 的配置。Spring Boot 使用"约定优先于配置（COC，Convention Over Configuration）"的理念，简单来说，Spring Boot 提供了针对企业应用开发各种场景的很多 spring-boot-starter 自动配置依赖模块，这些模块都基于"开箱即用"的原则，使得企业应用开发中各种场景的 Spring 应用更加快速和高效。

Spring Boot 是开发者和 Spring 框架的中间层，帮助开发者统筹管理应用的配置，提供基于实际开发中常见配置的默认处理（即约定优先于配置），简化应用的开发和运维；总体来说，Spring Boot 的目的就是为了对 Java Web 的开发进行"简化"和"加速"，简化开发过程中引入或启动相关 Spring 功能的配置。这样带来的好处就是降低开发人员对于框架的关注度，可以把更多的精力放在自己的业务代码上。

同时随着微服务概念的推广和实践，Spring Boot 的精简理念又使其成为 Java 微服务开发的不二之选，也可以说，Spring Boot 是最适合微服务的 Java Web 框架。关于微服务的更多知识请参考疯狂软件系列图书中的《疯狂 Spring Cloud 微服务架构实战》。

现如今，Spring Boot 已经在蓬勃发展的快速应用开发领域成为领导者。

1.2.2 Spring Boot 解决的问题

Spring Boot 的出现带来了以下优点。
- **使编码变得简单**：推荐使用注解。
- **使配置变得简单**：自动配置、快速构建项目、快速集成新技术的能力。
- **使部署变得简单**：内嵌 Tomcat、Jetty 等 Web 容器。
- **使监控变得简单**：自带项目监控。

1.2.3 Spring Boot 的主要特性

Spring Boot 的主要特点如下。
- Spring Boot 是伴随着 Spring 4.0 诞生的，继承了 Spring 框架原有的优秀基因。
- 遵循"约定优先于配置"的原则，使用 Spring Boot 只需很少的配置，大部分的时候直接使用默认的配置即可。
- 对主流开发框架无配置集成，自动整合第三方框架。
- 可独立运行 Spring 项目，Spring Boot 可以以 jar 包的形式独立运行。使用 java –jar 命令或者在项目的主程序中执行 main 函数就可以成功运行项目。
- 内嵌 Servlet 容器，可以选择内嵌 Tomcat、Jetty 等 Web 容器，无须以 war 包形式部署项目。
- 提供 starter 简化 Maven 配置，基本上可以做到自动化配置，高度封装，开箱即用。
- Spring Boot 会根据项目依赖来自动配置 Spring 框架，极大减少了项目所使用的配置。
- Spring Boot 提供了准生产环境的应用监控。
- 无代码生成和 XML 配置，纯 Java 的配置方式，很简单，很方便。
- 分布式开发，与 Spring Cloud 的微服务无缝结合。

1.2.4 Spring Boot 2.0 的重要改变

- 基于 Spring 5 构建，Spring 5 的新特性都可以在 Spring Boot 2.0 中使用。
- 为各种组件的响应式编程提供了简化配置，如：Reactive Spring Data、Reactive Spring Security 等。
- 要求 Java 版本必须是 Java 8 或更高版本，支持最新的 Java 9。
- 要求 Gradle 4 或更高版本、Maven 3.2 或更高版本。
- 要求 Tomcat 8 或更高版本，Hibernate 5.2 或更高版本，Thymeleaf 3 或更高版本。

1.3 "开箱即用"的依赖模块

Spring Boot 提供了针对企业应用开发各种场景的很多 spring-boot-starter 自动配置依赖模块，它们都约定以 spring-boot-starter-作为命名的前缀，并且都位于 org.springframework.boot 包或者命名空间下。

访问 http://start.spring.io，页面如图 1.2 所示。

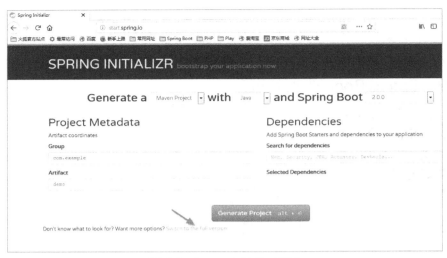

图 1.2　Spring Initializr 页面

单击图 1.2 所示的"Switch to the full version"超链接，可以看到展开的页面上显示的 Spring Boot 默认支持和提供了大约 80 个自动配置依赖模块。如此之多的依赖模块，开发者在实际项目中也不一定都会用到。此处重点讲解几个最常用的 spring-boot-starter 模块，其实所有的 spring-boot-starter 模块的使用都大同小异，读者可以之后在工作中灵活使用。

所有的 spring-boot-starter 模块都有约定的默认配置，但是允许开发者调整这些默认的配置用以改变默认的配置行为，这就是所谓的"约定优先于配置"。

简单来讲，Spring Boot 的配置主要可以分为以下几类：
- 命令行参数
- 系统环境变量
- 位于文件系统中的配置文件
- 位于 classpath 中的配置文件
- 固化到代码中的配置

以上几种方式按照优先级从高到低排列，高优先级方式提供的配置项会覆盖或者优先生效，比如通过命令行参数传入的配置项会覆盖通过环境变量传入的同一配置项。

实际项目开发中最常用的配置是配置文件，不管是位于文件系统还是位于 classpath，Spring Boot 应用默认的配置文件名叫作 application.properties，可以放在项目的 src/main/resources 目录下或者在类路径的 /config 目录下。

▶▶ 1.3.1　日志依赖模块 spring-boot-starter-logging

Java 的日志系统多种多样，有默认的 java.util 提供的日志支持，有 log4j、log4j2、commons logging 等日志框架。可以通过如下配置在 Maven 中添加 spring-boot-starter-logging 依赖模块。

```
<dependency>
    <groupId>org.springframework.boot</groupId>
    <artifactId>spring-boot-starter-logging</artifactId>
</dependency>
```

Spring Boot 将自动使用 logback 作为项目的日志框架。Spring Boot 为开发者提供了很多默认的日志配置，所以，只要将 spring-boot-starter-logging 依赖模块添加到当前项目，即可以"开箱即用"，不需要做任何多余的配置。

如果要对 Spring Boot 默认提供的日志设置做调整，则可以遵循 logback 的约定，使用自己定制的 logback.xml 日志文件，logback.xml 日志文件建议放在项目的 src/main/resources 目录下，然后在 application.properties 中指定。

```
logging.config = logback.xml
```

如果更习惯使用 log4j2，只需要采用类似的方式将对应的 spring-boot-starter 依赖模块加到 Maven 中。

```xml
<dependency>
    <groupId>org.springframework.boot</groupId>
    <artifactId>spring-boot-starter-log4j2</artifactId>
</dependency>
```

▶▶ 1.3.2 Web 开发依赖模块 spring-boot-starter-web

在互联网时代，大部分项目都是基于 Web 开发的，Spring 框架的使用者几乎都会使用 Spring MVC 来开发 Web 项目。为了帮助开发者简化快速搭建过程并开发 Web 项目，Spring Boot 提供了 spring-boot-starter-web 自动配置依赖模块。

可以通过如下配置在 Maven 中添加 spring-boot-starter-web 依赖模块。

```xml
<dependency>
    <groupId>org.springframework.boot</groupId>
    <artifactId>spring-boot-starter-web</artifactId>
</dependency>
```

这样，就得到了一个可以直接执行的 Web 项目，运行当前项目的 App 的 main 方法就可以直接启动一个使用了嵌入式 Tomcat 服务器的 Web 应用。只不过，在还没有提供任何服务 Web 请求的 Controller 时，访问任何路径都会返回一个 Spring Boot 默认提供的错误请求页面（称为 Whitelabel Error Page），如图 1.3 所示。

图 1.3 Spring Boot 默认错误页面

可以在当前项目下新建一个 Spring MVC 的 Controller 控制器，代码如下。

```java
@RestController
public class IndexController {
    // 映射"/"请求
    @RequestMapping("/")
    public String index(){
        return "Hello Spring Boot!";
    }
}
```

重新运行 App 的 main 方法并访问 http://127.0.0.1:8080，将正常显示 Controller 返回的信息"Hello Spring Boot!"，一个简单的 Web 项目就完成了。

 提示
> 本节重点讲解 spring-boot-starter-web 依赖模块的作用，关于 Spring Boot 项目的搭建、Spring MVC 的 Controller 控制器的使用，1.4 节会重点介绍。

Spring Boot 使用 spring-boot-starter-web 依赖模块开发 Web 项目，非常简单、方便。但是，在简单的背后，其实有很多的"约定"，只有充分了解这些"约定"，才能更好地使用 spring-boot-starter-web 依赖模块。

（1）项目结构层面的约定

在传统的 Java Web 项目中，所有的静态文件和页面都放在 WebContent 目录下，而 Spring Boot 项目的静态文件和页面统一放在 src/main/resources 目录对应的子目录下。src/main/resources/static 目录用于存放各类静态资源文件，比如 css、js 和 image 等；src/main/resources/templates 目录用于存放页面模板文件，比如 html 和 jsp 等。

（2）Spring MVC 框架层面的约定

spring-boot-starter-web 依赖模块默认自动配置一些 Spring MVC 必要的组件。

- 将 ViewResolver 自动注册到 Spring 容器。
- 将 Converter 和 Formatter 等 bean 自动注册到 Spring 容器。
- 将对 Web 请求的支持和相应的类型转换的 HttpMessageConverter 自动注册到 Spring 容器。
- 将 MessageCodesResolver 自动注册到 Spring 容器。

（3）嵌入式 Web 容器层面的约定

spring-boot-starter-web 依赖模块默认使用嵌入式 Tomcat 作为 Web 容器对外提供服务，默认使用 8080 端口对外监听和提供服务。

如果在开发中不想使用默认的嵌入式 Tomcat，可以引入 jetty 作为替代方案。

```xml
<dependency>
    <groupId>org.springframework.boot</groupId>
    <artifactId>spring-boot-starter-jetty</artifactId>
</dependency>
```

或者使用 undertow 作为替代方案。

```xml
<dependency>
    <groupId>org.springframework.boot</groupId>
    <artifactId>spring-boot-starter-undertow</artifactId>
</dependency>
```

如果不想使用默认的 8080 端口，可以通过更改 application.properties 配置文件中的 server.port 使用自己指定的端口。

```
server.port=8088
```

Spring Boot 提供了大量的 spring-boot-starter- 依赖模块，后面的章节会详细讲解。

1.4 开发第一个 Spring Boot 应用

本书重点介绍 Spring Boot，使用 Eclipse 作为 Java IDE 开发工具，使用 Maven 作为项目构建工具，关于 Eclipse 和 Maven 的详细安装和使用、Spring 的使用请参考疯狂软件系列图书中的《轻量级 Java EE 企业应用实战（第 5 版）：Struts 2+Spring 5+Hibernate 5/JPA 2 整合开发》，关于 Spring MVC 的使用请参考疯狂软件系列图书中的《Spring+MyBatis 企业应用实战》。

1.4.1 下载和安装 Maven

下载和安装 Maven 请按如下步骤进行。

① 登录 http://maven.apache.org/download.cgi 站点下载 Maven 最新版，本书成书之时，Maven 的最新稳定版是 3.5.0，建议下载该版本。

虽然 Maven 是基于 Java 的生成工具，具有平台无关的特性，但考虑到解压缩的方便性，通常建议 Windows 平台下载*.zip 压缩包，而 Linux 平台则下载*.gz 压缩包。

②将下载到的压缩文件解压缩到任意路径，此处将其解压缩到 D:\路径下，解压缩后生成文件夹 apache-maven-3.5.0。解压缩后看到如下文件结构。

- bin：保存 Maven 的可执行命令。其中 mvn 和 mvn.bat 就是执行 Maven 工具的命令。
- boot：该目录只包含一个 plexus-classworlds-2.5.2.jar。plexus-classworlds 是一个类加载器框架，与默认的 Java 类加载器相比，它提供了更丰富的语法以方便配置，Maven 使用该框架加载自己的类库。通常无须理会该文件。
- conf：保存 Maven 配置文件的目录，该目录包含 settings.xml 文件，该文件用于设置 Maven 的全局行为。
- lib：该目录包含了所有 Maven 运行时需要的类库，Maven 本身是分模块开发的，因此用户能看到诸如 maven-core-3.5.0.jar、maven-repository-metadata-3.5.0.jar 等文件。此外，还包含 Maven 所依赖的第三方类库。
- LICENSE、NOTICE、README.txt 等说明性文档。

③配置 Maven 本地资源库。

在 apache-maven-3.5.0 文件夹下新建文件夹 repository，用于充当本地资源库。打开 apache-maven-3.5.0\conf\setting.xml 文件，在<settings xmlns/>元素下增加

```
<localRepository>D:/apache-maven-3.5.0/repository</localRepository>
```

<localRepository>元素的内容是一个路径字符串，该路径用于设置 Maven 的本地资源库的路径。如果用户不设置该参数，Maven 本地资源库默认保存在用户 Home 目录的 m2/repository 路径下。考虑到 Windows 有可能需要重装、恢复系统，因此建议该 Maven 本地资源库设置到其他路径。

资源库是 Maven 的一个重要概念，Maven 构建项目所使用的插件、第三方依赖库都集中存放在本地资源库中。

1.4.2 Eclipse 集成 Maven

打开 Eclipse，选择 Window→Preferences→Maven→User Settings，如图 1.4 所示。

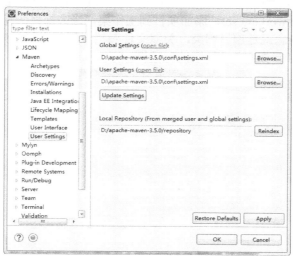

图 1.4　Eclipse 集成 Maven

在 User Settings 窗格中包含如下选项。

- Global Settings：全局设置，此处选择 Maven 的 conf 目录下的 settings.xml。

- User Settings：用户设置，此处选择 Maven 的 conf 目录下的 settings.xml。
- Local Repository：Maven 的本地资源库，即创建的 repository 目录。

1.4.3 示例：第一个 Spring Boot 应用

1.4.3.1 创建一个新的 Maven 项目

打开 Eclipse，选择 File→New→Others→Maven→Maven Project 命令，打开如图 1.5 所示的对话框。

图 1.5 创建新的 Maven 项目

选择项目路径：

- Create a simple project (skip archetype selection)，创建一个简单项目（跳过原型选择）。如果选择该选项将直接跳过 Maven 的项目原型（模板）选择，建议不要勾选，可以使用内置的模板。
- Use default Workspace location，使用默认的工作区间，勾选后，建立的项目将放在默认工作区间，如不选则单击 Browse（浏览）按钮，选择一个工作区间。
- Add project(s) to working set，添加项目到工作集，选中则将新建的项目放入工作集，这里的工作集的概念就是项目归类，类似文件归档一样，方便区分。此选项可选可不选。

单击 Next 按钮，打开如图 1.6 所示的对话框。

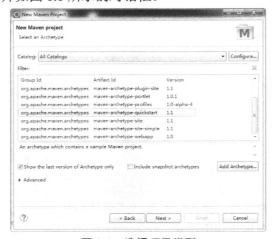

图 1.6 选择项目类型

选择项目类型：

通常选择建立 maven-archetype-quickstart（非 Web 项目）项目模型或者 maven-archetype-webapp（Web 项目）项目模型，这两种是比较常用的模型。

单击 Next 按钮，打开如图 1.7 所示的对话框。

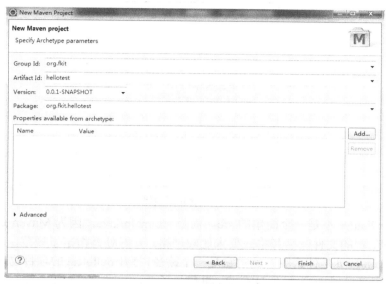

图 1.7　填写项目参数

填写项目参数：

➢ Group Id，项目组织唯一的标识符，比如：org.***或者 com.****，本例输入"org.fkit"。
➢ Artifact Id，项目的名称，本例为"hellotest"。
➢ Version，当前版本，此处选择默认项。
➢ Package，默认包结构。

单击 Finish 按钮，会创建一个 maven-archetype-quickstart 项目（即非 Web 项目）。第一次创建 Maven 项目时，会下载必需的 jar 包，如图 1.8 所示。

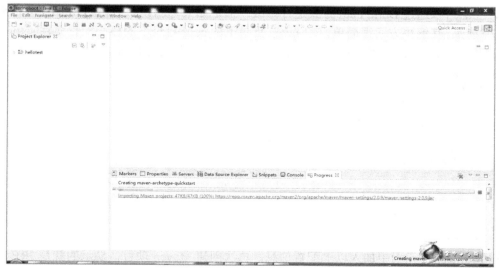

图 1.8　自动下载所需 jar 文件

根据网络速度，下载所需的 jar 文件可能需要几分钟或更长时间。

1.4.3.2 Maven 项目结构

hellotest 项目创建成功后，在 Package Explorer 里面看到的项目目录结构如图 1.9 所示。

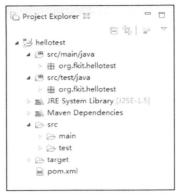

图 1.9 Maven 项目结构

（1）Source Folder 不是一个简单的 src，而是 src/main/java，因为 Maven 是一种强约束的工程类型。它对工程的文件命名和格式要求比较严格。其好处是指定了规范，方便代码的移植和理解。而 src/main/java 是什么呢？其实它是一个路径，打开 hellotest 的项目物理地址会发现，是一个 src 文件夹包含了一个 main 文件夹，再包含了 java 文件夹。java 文件夹下有两个目录。

➢ src/main/java：这个目录下储存主要的 Java 代码。
➢ src/test/java：储存测试用的类，比如 JUnit 的测试一般就放在这个目录下面。

（2）target 文件夹是源码编译后生成的 class 文件存放的地方（如果是一个 Web 应用，还有其他的信息也在编译打包之后放在 target 里面）。target 文件夹下有两个目录。

➢ classes：存放项目代码编译后生成的 class 文件。
➢ test-classes：存放测试代码编译后生成的 class 文件。

（3）细心的读者可能会发现项目中的 JRE System Library [J2SE-1.5]，当前使用的环境明明是 Java 8，为什么会出现[J2SE-1.5]呢？因为到目前为止，所有 Maven 版本的默认 JDK 版本都是 1.5，当然可以通过 Build Path 方式进行修改。但是如果每创建一个新的 Maven 项目都要修改会非常麻烦。其解决方案很简单，在 Maven 目录下的 conf 目录下的核心配置文件 settings.xml 中加入配置信息，默认使用 JDK 1.8 即可。settings.xml 文件修改内容如下：

```xml
<profiles>
   <profile>
      <id>jdk-1.8</id>
      <activation>
         <activeByDefault>true</activeByDefault>
         <jdk>1.8</jdk>
      </activation>
      <properties>
         <maven.compiler.source>1.8</maven.compiler.source>
         <maven.compiler.target>1.8</maven.compiler.target>
         <maven.compiler.compilerVersion>1.8</maven.compiler.compilerVersion>
      </properties>
   </profile>
</profiles>
```

在<profiles.../>元素中加入以上配置信息，再次创建 Maven 项目，显示的就是 JRE System Library [J2SE-1.8]了。

（4）pom.xml 文件是 Maven 的基础配置文件，也是项目中最重要的文件，被称为项目对象模型（Project Object Model）描述文件。Maven 使用一种被称为项目对象模型的方式来管理项目，POM 用于描述如下这些问题：该项目是什么类型的？该项目的名称是什么？该项目的构建能自定义吗？Maven 使用 pom.xml 文件来描述项目对象模型。因此，pom.xml 并不是简单的生成文件，而是一种项目对象模型的描述文件。pom.xml 文件内容如下：

```xml
<project xmlns="http://maven.apache.org/POM/4.0.0" xmlns:xsi="http://www.w3.org/2001/XMLSchema-instance"
    xsi:schemaLocation="http://maven.apache.org/POM/4.0.0 http://maven.apache.org/xsd/maven-4.0.0.xsd">
    <modelVersion>4.0.0</modelVersion>
    <groupId>org.fkit.springboot</groupId>
    <artifactId>hellotest</artifactId>
    <version>0.0.1-SNAPSHOT</version>
    <packaging>jar</packaging>
    <name>hellotest</name>
    <url>http://maven.apache.org</url>
    <properties>
        <project.build.sourceEncoding>UTF-8</project.build.sourceEncoding>
    </properties>
    <dependencies>
        <dependency>
            <groupId>junit</groupId>
            <artifactId>junit</artifactId>
            <version>3.8.1</version>
            <scope>test</scope>
        </dependency>
    </dependencies>
</project>
```

如上所示的 pom.xml 文件将是 Maven 项目中最简单的，一般来说，实际的 pom.xml 文件将比这个文件更复杂：它会定义多个依赖、定义额外的插件等。上面这个 pom.xml 中的 <groupId.../>、<artifactId.../>、<version.../>、<packaging.../>定义了该项目的唯一标识（几乎所有的 pom.xml 文件都需要它们），这个唯一标识被称为 Maven 坐标（coordinate）。<name.../> 和<url.../>元素只是 pom.xml 提供的描述性元素，用于提供可阅读的名字。<properties.../>元素用于定义全局属性，<project.build.sourceEncoding.../>元素定义的是项目的字符编码。<dependencies.../>元素定义了一个单独的测试范围的依赖，这表明该项目的测试依赖于 JUnit 测试框架。

接下来，修改 pom.xml 文件。在<url.../>元素之后增加<parent.../>元素信息。

```xml
<parent>
    <groupId>org.springframework.boot</groupId>
    <artifactId>spring-boot-starter-parent</artifactId>
    <version>2.0.0.RELEASE</version>
</parent>
```

构建基于 Spring Boot 的应用首先必须要将<parent.../>元素设置为 Spring Boot 的 spring-boot-starter-parent，spring-boot-starter-parent 是 Spring Boot 的核心启动器，包含了自动配置、日志和 YAML 等大量默认的配置，大大简化了开发工作。<version.../>元素建议使用最新的 RELEASE 版本，之后的 Spring Boot 模块（如 spring-boot-starter-web 模块）都会自动选择最合适的版本进行添加。本书使用的是写作时最新的 2.0.0.RELEASE，读者测试本书示例代码时建议使用 2.0.0 或更高版本。

在<dependencies.../>元素中增加一个<dependency.../>元素内容。

```xml
<dependency>
    <groupId>org.springframework.boot</groupId>
```

```xml
        <artifactId>spring-boot-starter-web</artifactId>
</dependency>
```

<dependency.../> 元素用于添加需要的 starter 模块，作为示例，此处只添加了 spring-boot-starter-web 模块。Spring Boot 中包含了很多 starter 模块，简单来说，每一个 starter 模块就是一系列的依赖包组合。例如 starter-web 模块，就是包含了 Spring Boot 预定义的一些 Web 开发的常用依赖包，支持全栈式 Web 开发，包括 Tomcat 和 spring-webmvc。由于指定了 spring-boot-starter-parent，所以此处的 web starter 模块不需要指定版本，Spring Boot 会自动选择最合适的版本进行添加。

修改完成之后的 pom.xml 文件内容如下。

程序清单：codes/01/hellotest/pom.xml

```xml
<project xmlns="http://maven.apache.org/POM/4.0.0" xmlns:xsi="http://www.w3.org/2001/XMLSchema-instance"
    xsi:schemaLocation="http://maven.apache.org/POM/4.0.0 http://maven.apache.org/xsd/maven-4.0.0.xsd">
    <modelVersion>4.0.0</modelVersion>
    <groupId>org.fkit.springboot</groupId>
    <artifactId>hellotest</artifactId>
    <version>0.0.1-SNAPSHOT</version>
    <packaging>jar</packaging>
    <name>hellotest</name>
    <url>http://maven.apache.org</url>

    <parent>
        <groupId>org.springframework.boot</groupId>
        <artifactId>spring-boot-starter-parent</artifactId>
        <version>2.0.0.RELEASE</version>
    </parent>

    <properties>
        <project.build.sourceEncoding>UTF-8</project.build.sourceEncoding>
        <project.reporting.outputEncoding>UTF-8</project.reporting.outputEncoding>
        <java.version>1.8</java.version>
    </properties>

    <dependencies>
        <dependency>
            <groupId>org.springframework.boot</groupId>
            <artifactId>spring-boot-starter-web</artifactId>
        </dependency>
        <dependency>
            <groupId>junit</groupId>
            <artifactId>junit</artifactId>
            <scope>test</scope>
        </dependency>
    </dependencies>
</project>
```

提示

pom.xml 文件修改保存后，Maven 会自动下载所需的所有 jar 文件。

1.4.3.3 编写测试代码

① 写一个简单的 Java 类 HelloController。

程序清单：codes/01/hellotest/org/fkit /hellotest/HelloController.java

```java
import org.springframework.web.bind.annotation.RequestMapping;
import org.springframework.web.bind.annotation.RestController;
```

```java
// RestController 相当于 SpringMVC 中的 @Controller + @ResponseBody
@RestController
public class HelloController {
    // 映射"/hello"请求
    @RequestMapping("/hello")
    public String hello(){
        return " Hello Spring Boot!";
    }
}
```

HelloController 类中使用的@RestController 注解是一个组合注解，相当于 Spring MVC 中的@Controller+@ResponseBody 合在一起的作用。表明这个请求返回字符串"hello"。

② 修改 Maven 默认的 App 类。

程序清单：codes/01/hellotest/org/fkit/hellotest/App.java

```java
import org.springframework.boot.SpringApplication;
import org.springframework.boot.autoconfigure.SpringBootApplication;
// @SpringBootApplication 指定这是一个 Spring Boot 的应用程序
@SpringBootApplication
public class App
{
    public static void main( String[] args )
    {
        // SpringApplication 用于从 main 方法启动 Spring 应用的类
        SpringApplication.run(App.class, args);
    }
}
```

App 类使用的@SpringBootApplication 注解指定这是一个 Spring Boot 的应用程序，该注解也是一个组合注解，相当于@Configuration + @EnableAutoConfiguration + @ComponentScan，细节在第 2 章重点讲解。SpringApplication 类用于从 main 方法启动 Spring 应用的类。此处直接调用静态的 run 方法。

③ 启动 Spring Boot 项目。

使用右键快捷菜单运行 main 方法。控制台信息如下。

```
  .   ____          _            __ _ _
 /\\ / ___'_ __ _ _(_)_ __  __ _ \ \ \ \
( ( )\___ | '_ | '_| | '_ \/ _` | \ \ \ \
 \\/  ___)| |_)| | | | | || (_| |  ) ) ) )
  '  |____| .__|_| |_|_| |_\__, | / / / /
 =========|_|==============|___/=/_/_/_/
 :: Spring Boot ::        (v2.0.0.RELEASE)

①2018-03-28 11:52:23.958  INFO 9188 --- [           main] org.fkit.hellotest.App                   : Starting App on PC-201610061142 with PID 9188 (C:\Users\Administrator\Desktop\hellotest\target\classes started by Administrator in C:\Users\Administrator\Desktop\hellotest)
②2018-03-28 11:52:23.962  INFO 9188 --- [           main] org.fkit.hellotest.App                   : No active profile set, falling back to default profiles: default
③2018-03-28 11:52:24.055  INFO 9188 --- [           main] ConfigServletWebServerApplicationContext : Refreshing org.springframework.boot.web.servlet.context.AnnotationConfigServletWebServerApplicationContext@4461c7e3: startup date [Wed Mar 28 11:52:24 CST 2018]; root of context hierarchy
④2018-03-28 11:52:25.698  INFO 9188 --- [           main] o.s.b.w.embedded.tomcat.TomcatWebServer  : Tomcat initialized with port(s): 8080 (http)
⑤2018-03-28 11:52:25.732  INFO 9188 --- [           main] o.apache.catalina.core.StandardService   : Starting service [Tomcat]
```

⑥2018-03-28 11:52:25.732 INFO 9188 --- [main] org.apache.catalina.core.StandardEngine : Starting Servlet Engine: Apache Tomcat/8.5.28
⑦2018-03-28 11:52:25.743 INFO 9188 --- [ost-startStop-1] o.a.catalina.core.AprLifecycleListener   : The APR based Apache Tomcat Native library which allows optimal performance in production environments was not found on the java.library.path: [D:\java\soft\jdk1.8.0_20_x64\bin;C:\Windows\Sun\Java\bin;C:\Windows\system32;C:\Windows;D:/java/soft/jdk1.8.0_20_x64/bin/../jre/bin/server;D:/java/soft/jdk1.8.0_20_x64/bin/../jre/bin;D:/java/soft/jdk1.8.0_20_x64/bin/../jre/lib/amd64;C:\Windows\system32;C:\Windows;C:\Windows\System32\Wbem;C:\Windows\System32\WindowsPowerShell\v1.0\;C:\Program Files\Intel\WiFi\bin\;C:\Program Files\Common Files\Intel\WirelessCommon\;C:\Program Files\Lenovo\Fingerprint Manager Pro\;C:\Program Files (x86)\NVIDIA Corporation\PhysX\Common;D:\java\soft\jdk1.8.0_20_x64\bin;C:\Program Files\Intel\WiFi\bin\;C:\Program Files\Common Files\Intel\WirelessCommon\;D:\java\soft\eclipse;.]
⑧2018-03-28 11:52:25.869 INFO 9188 --- [ost-startStop-1] o.a.c.c.C.[Tomcat].[localhost].[/]   : Initializing Spring embedded WebApplicationContext
⑨2018-03-28 11:52:25.870 INFO 9188 --- [ost-startStop-1] o.s.web.context.ContextLoader   : Root WebApplicationContext: initialization completed in 1819 ms
⑩2018-03-28 11:52:26.009 INFO 9188 --- [ost-startStop-1] o.s.b.w.servlet.ServletRegistrationBean : Servlet dispatcherServlet mapped to [/]
⑪2018-03-28 11:52:26.013 INFO 9188 --- [ost-startStop-1] o.s.b.w.servlet.FilterRegistrationBean   : Mapping filter: 'characterEncodingFilter' to: [/*]
⑫2018-03-28 11:52:26.013 INFO 9188 --- [ost-startStop-1] o.s.b.w.servlet.FilterRegistrationBean   : Mapping filter: 'hiddenHttpMethodFilter' to: [/*]
⑬2018-03-28 11:52:26.013 INFO 9188 --- [ost-startStop-1] o.s.b.w.servlet.FilterRegistrationBean   : Mapping filter: 'httpPutFormContentFilter' to: [/*]
⑭2018-03-28 11:52:26.013 INFO 9188 --- [ost-startStop-1] o.s.b.w.servlet.FilterRegistrationBean   : Mapping filter: 'requestContextFilter' to: [/*]
⑮2018-03-28 11:52:26.346 INFO 9188 --- [main] s.w.s.m.m.a.RequestMappingHandlerAdapter : Looking for @ControllerAdvice: org.springframework.boot.web.servlet.context.AnnotationConfigServletWebServerApplicationContext@4461c7e3: startup date [Wed Mar 28 11:52:24 CST 2018]; root of context hierarchy
⑯2018-03-28 11:52:26.436 INFO 9188 --- [main] s.w.s.m.m.a.RequestMappingHandlerMapping : Mapped "{[/hello]}" onto public java.lang.String org.fkit.hellotest.HelloController.hello()
⑰2018-03-28 11:52:26.441 INFO 9188 --- [main] s.w.s.m.m.a.RequestMappingHandlerMapping : Mapped "{[/error]}" onto public org.springframework.http.ResponseEntity<java.util.Map<java.lang.String,java.lang.Object>> org.springframework.boot.autoconfigure.web.servlet.error.BasicErrorController.error(javax.servlet.http.HttpServletRequest)
⑱2018-03-28 11:52:26.442 INFO 9188 --- [main] s.w.s.m.m.a.RequestMappingHandlerMapping : Mapped "{[/error],produces=[text/html]}" onto public org.springframework.web.servlet.ModelAndView org.springframework.boot.autoconfigure.web.servlet.error.BasicErrorController.errorHtml(javax.servlet.http.HttpServletRequest,javax.servlet.http.HttpServletResponse)
⑲2018-03-28 11:52:26.479 INFO 9188 --- [main] o.s.w.s.handler.SimpleUrlHandlerMapping : Mapped URL path [/webjars/**] onto handler of type [class org.springframework.web.servlet.resource.ResourceHttpRequestHandler]
⑳2018-03-28 11:52:26.480 INFO 9188 --- [main] o.s.w.s.handler.SimpleUrlHandlerMapping : Mapped URL path [/**] onto handler of type [class org.springframework.web.servlet.resource.ResourceHttpRequestHandler]
㉑2018-03-28 11:52:26.517 INFO 9188 --- [main]

```
o.s.w.s.handler.SimpleUrlHandlerMapping  : Mapped URL path [/**/favicon.ico] onto
handler of type [class
org.springframework.web.servlet.resource.ResourceHttpRequestHandler]
    ㉒2018-03-28 11:52:26.681  INFO 9188 --- [           main]
o.s.j.e.a.AnnotationMBeanExporter        : Registering beans for JMX exposure on startup
    ㉓2018-03-28 11:52:26.730  INFO 9188 --- [           main]
o.s.b.w.embedded.tomcat.TomcatWebServer  : Tomcat started on port(s): 8080 (http) with
context path ''
    ㉔2018-03-28 11:52:26.733  INFO 9188 --- [           main]
org.fkit.hellotest.App                   : Started App in 3.114 seconds (JVM running for
3.486)
```

从控制台的信息中,可以看到 Spring Boot 的版本、Apache Tomcat 的版本、监听的端口、Spring 的 WebApplicationContext 创建信息等日志信息。有趣的是,在日志信息上方看到了一个 Spring 的图形,该图形还可以自行定制。

深入分析控制台启动信息:

第 1 行,启动 App。
第 2 行,查找 active profile,若无,则设为 default。
第 3 行,刷新上下文。
第 4 行,初始化 Tomcat,设置端口 8080,设置访问方式为 http。
第 5 行,启动 Tomcat 服务。
第 6 行,启动 Servlet 引擎。
第 7 行,Spring 内嵌的 WebApplicationContext,初始化开始。
第 8 行,Spring 内嵌的 WebApplicationContext,初始化完成。
第 10 行,映射 servlet,将 dispatcherServlet 映射到 [/]。
第 11 行,映射 filter,将 characterEncodingFilter 映射到 [/*]。
第 12 行,映射 filter,将 hiddenHttpMethodFilter 映射到 [/*]。
第 13 行,映射 filter,将 httpPutFormContentFilter 映射到 [/*]。
第 14 行,映射 filter,将 requestContextFilter 映射到 [/*]。
第 15 行,查找 @ControllerAdvice。
第 16 行,路径 "{[/hello]}" 映射到 org.fkit.springboot.hellotest.HelloController.hello()。
第 17 行,路径 "{[/error]}" 映射到 org.springframework.boot.autoconfigure.web. BasicErrorController.error(javax.servlet.http.HttpServletRequest)。
第 18 行,将路径 "{[/error],produces=[text/html]}" 映射到 org.springframework.web.servlet.ModelAndView org.springframework.boot.autoconfigure.web.BasicErrorController.errorHtml (javax. servlet. http.HttpServletRequest,javax.servlet.http.HttpServletResponse)。
第 23 行,Tomcat 启动完毕。
第 24 行,App 启动耗费的时间。

从上面的启动信息中可以明显看到 Tomcat 启动的过程、Spring MVC 的加载过程,包括 dispatcherServlet 等。

④ 测试应用。

Spring Boot 项目启动后,默认访问地址为:http://localhost:8080/,按照之前的 Web 项目习惯,读者可能会问,怎么没有项目路径?这就是 Spring Boot 的默认设置了,将项目路径直接设为根路径。

在浏览器中输入 URL 来测试应用。

```
http://localhost:8080/hello
```

看到如图 1.10 所示页面，表示 Spring Boot 项目访问成功。

图 1.10　访问 Spring Boot 项目

 ## 1.5　本章小结

本章主要介绍了 Spring 框架，Spring Boot 的核心功能，Maven 的下载和安装，Eclipse 集成 Maven 开发 Spring Boot 项目，使用 Spring Boot 可以不用或者只需很少的 Spring 配置就可以让企业项目快速运行起来。

第 2 章将重点介绍 Spring Boot 的核心内容。

第 2 章
Spring Boot 核心

本章要点

- Spring Boot 的核心注解
- Spring Boot 的基本配置
- Spring Boot 的自动配置原理

在第 1 章中，通过一个小的入门案例，已经让我们看到了 Spring Boot 的强大之处。本章将详细介绍 Spring Boot 的核心注解、基本配置和自动配置的原理。

2.1 Spring Boot 启动类与核心注解@SpringBootApplication

Spring Boot 的项目一般都会有注解*Application 标注的入口类，入口类中会有一个 main 方法，main 方法是一个标准的 Java 应用程序的入口方法，可以直接启动。

@SpringBootApplication 注解是 Spring Boot 的核心注解，用此注解标注的入口类是应用的启动类，通常会在启动类的 main 方法中通过 SpringApplication.run(App.class, args) 来启动 Spring Boot 应用项目。

@SpringBootApplication 其实是一个组合注解，源代码如下。

程序清单：org/springframework/boot/autoconfigure/SpringBootApplication.java

```java
@Target(ElementType.TYPE)
@Retention(RetentionPolicy.RUNTIME)
@Documented
@Inherited
@SpringBootConfiguration
@EnableAutoConfiguration
@ComponentScan(excludeFilters = {
        @Filter(type = FilterType.CUSTOM, classes = TypeExcludeFilter.class),
        @Filter(type = FilterType.CUSTOM, classes = AutoConfigurationExcludeFilter.class) })
public @interface SpringBootApplication {
@AliasFor(annotation = EnableAutoConfiguration.class, attribute = "exclude")
    Class<?>[] exclude() default {};

    @AliasFor(annotation = EnableAutoConfiguration.class, attribute = "excludeName")
    String[] excludeName() default {};

    @AliasFor(annotation = ComponentScan.class, attribute = "basePackages")
    String[] scanBasePackages() default {};

    @AliasFor(annotation = ComponentScan.class, attribute = "basePackageClasses")
    Class<?>[] scanBasePackageClasses() default {};
}
```

@SpringBootApplication 注解主要组合了以下注解。

（1）@SpringBootConfiguration：这是 Spring Boot 项目的配置注解，也是一个组合注解，源代码如下。

程序清单：org/springframewor/boot/SpringBootConfiguration.java

```java
@Target(ElementType.TYPE)
@Retention(RetentionPolicy.RUNTIME)
@Documented
@Configuration
public @interface SpringBootConfiguration {

}
```

在 Spring Boot 项目中推荐使用@SpringBootConfiguration 注解来替代@Configuration 注解。

（2）@EnableAutoConfiguration：启动自动配置，该注解会让 Spring Boot 根据当前项目所依赖的 jar 包自动配置项目的相关配置项。

例如，当在 Spring Boot 项目的 pom.xml 文件中配置了如下 spring-boot-starter-web 依赖。

```xml
<dependency>
    <groupId>org.springframework.boot</groupId>
    <artifactId>spring-boot-starter-web</artifactId>
</dependency>
```

项目就会自动添加 Tomcat 和 Spring MVC 的依赖，同时 Spring Boot 会对 Tomcat 和 Spring MVC 进行配置项的自动配置，打开 pom.xml，选择 Dependency Hierarchy 页面查看，如图 2-1 所示。

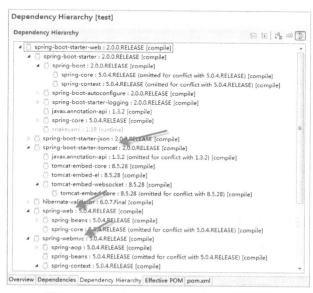

图 2.1　spring-boot-starter-web 自动配置图

图 2.1 显示的是当前项目中 Spring Boot 的所有依赖，如果在项目中添加 spring-boot-starter-data-jpa 依赖，Spring Boot 就会自动进行 spring-boot-starter-data-jpa 依赖的相关配置，读者可以自行添加依赖，然后进行观察。

（3）@ComponentScan：扫描配置，Spring Boot 默认会扫描@SpringBootApplication 所在类的同级包以及它的子包，所以建议将@SpringBootApplication 修饰的入口类放置在项目包下（Group Id+Artifact Id），这样做的好处是，可以保证 Spring Boot 项目自动扫描到项目所有的包。

2.2　Spring Boot 基本配置介绍

2.2.1　关闭某个自动配置

通过上述@SpringBootApplication 下的@EnableAutoConfiguration 可知，Spring Boot 会根据项目中的 jar 包依赖，自动做出配置，Spring Boot 支持的部分自动配置如下（非常多），如图 2.2 所示。

假如不需要 Spring Boot 自动配置，想关闭某一项的自动配置，该如何设置呢？

例如不想自动配置 Redis，想自己手动配置，通过查看@SpringBootApplication 的源码可以看出，关闭特定的自动配置应该使用@SpringBootApplication 下的 exclude 参数，现以关闭 Redis 自动配置为例：

```
@SpringBootApplication(exclude={RedisAutoConfiguration.class})
```

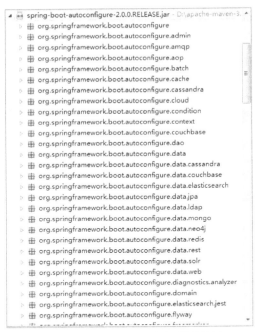

图 2.2　Spring Boot 的自动配置项

▶▶ 2.2.2　定制启动 banner

在启动 Spring Boot 项目的时候，我们在控制台看到了如图 2.3 所示的默认启动图案。

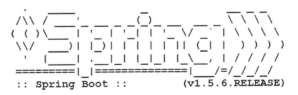

图 2.3　默认启动图案

如果想自己指定启动的图案应该如何配置呢？

① 在浏览器中打开网站 http://patorjk.com/software/taag，如图 2.4 所示。

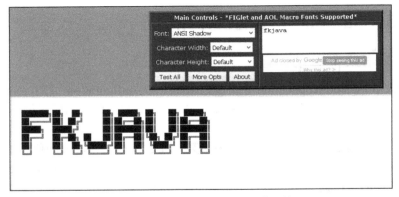

图 2.4　自定义 banner 的网站示范

② 在第 1 步所示范的网站上单击左下方的"Select & Copy"按钮，将自定义的 banner 图

案进行复制，然后新建一个 banner.txt 文件，将复制好的图案写入 banner.txt 文件中。

③ 将 banner.txt 文件放到项目的 src/main/resources 目录下，如图 2.5 所示。

图 2.5　banner.txt 文件的放置位置

④ 重新启动程序，效果如图 2.6 所示。

图 2.6　banner 的启动效果

2.2.3　应用的全局配置文件

可以在 Spring Boot 项目的 src/main/resources 目录下或者在类路径的/config 目录下创建一个全局的配置文件 application.properties 或者是后缀为.yml 的 application.yml 文件，用于修改 Spring Boot 项目的默认配置值。例如修改项目的默认端口，或者进入 DispatcherServlet 的请求地址规则等。

通常，在实际开发中我们习惯使用 application.properties 文件作为应用的全局配置文件，一般放到 src/main/resources 目录下。

例如，在 src/main/resources 目录下创建一个名为 application.properties 的文件，配置内容如下：

```
server.port=9999
server.servlet-path=*.action
```

（1）其中，server.port 参数用于将 Spring Boot 项目的默认端口改为 9999。启动应用，端口修改后如图 2.7 所示。

（2）server.servlet-path 参数用于将进入 DispatcherServlet 的规则修改为：*.action，测试效果如图 2.8 所示。

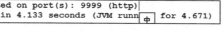

图 2.7　修改项目端口图

图 2.8　测试访问项目

从上面的参数配置可以看出，Spring Boot 支持很多参数的配置与参数值的修改，关于其他配置参数的详细说明和描述可以查看官方的文档说明：http://docs.spring.io/spring-boot/docs/

1.5.6.RELEASE/reference/htmlsingle/#common-application-properties。

2.2.4 Spring Boot 的依赖模块

Spring Boot 提供了很多简化项目开发的"开箱即用"的依赖模块，例如在项目中使用的 pom.xml 文件中的配置。

```xml
<dependency>
    <groupId>org.springframework.boot</groupId>
    <artifactId>spring-boot-starter-web</artifactId>
</dependency>
```

Spring Boot 会自动关联 Web 开发相关的依赖，如 Tomcat 以及 spring-webmvc 等，进而对 Web 开发进行支持，同时相关技术的配置也将实现自动配置，程序员即避开烦琐的配置文件了。除此之外，官方还提供了如下依赖模块。

- spring-boot-starter：这是 Spring Boot 的核心启动器，包含了自动配置、日志和 YAML 文件的支持。
- spring-boot-starter-activemq：为 JMS 使用 Apache ActiveMQ。ActiveMQ 是 Apache 出品的最流行、能力强劲的开源消息总线。
- spring-boot-starter-amqp：通过 spring-rabbit 来支持 AMQP 协议（Advanced Message Queuing Protocol）。
- spring-boot-starter-aop：支持面向切面的编程（AOP），包括 spring-aop 和 AspectJ。
- spring-boot-starter-artemis：通过 Apache Artemis 支持 JMS 的 API（Java Message Service API）。
- spring-boot-starter-batch：支持 Spring Batch，包括 HSQLDB 数据库。
- spring-boot-starter-cache：支持 Spring 的 Cache 抽象。
- spring-boot-starter-cloud-connectors：支持 Spring Cloud Connectors，简化了在像 Cloud Foundry 或 Heroku 这样的云平台上的连接服务。
- spring-boot-starter-data-cassandra：使用 Cassandra 分布式数据库、Spring Data Cassandra。Apache Cassandra 是一套开源分布式 NoSQL 数据库系统。
- spring-boot-starter-data-couchbase：使用 Couchbase 文件存储数据库、Spring Data Couchbase。Spring Data 是一个用于简化数据库访问并支持云服务的开源框架。
- spring-boot-starter-data-elasticsearch：支持 ElasticSearch 搜索和分析引擎，包括 spring-data-elasticsearch。
- spring-boot-starter-data-gemfire：支持 GemFire 分布式数据存储，包括 spring-data-gemfire。
- spring-boot-starter-data-jpa：支持 JPA（Java Persistence API），包括 spring-data-jpa、spring-orm 和 Hibernate。
- spring-boot-starter-data-ldap：支持 Spring Data LDAP。
- spring-boot-starter-data-mongodb：支持 MongoDB 数据库，包括 spring-data-mongodb。
- spring-boot-starter-data-neo4j：使用 Neo4j 图形数据库、Spring Data Neo4j。Neo4j 是一个高性能的 NoSQL 图形数据库，它将结构化数据存储在网络上而不是表中。
- spring-boot-starter-redis：支持 Redis 键值存储数据库，包括 spring-redis。
- spring-boot-starter-data-rest：通过 spring-data-rest-webmvc，支持通过 REST 暴露 Spring Data 数据仓库。
- spring-boot-starter-data-solr：支持 Apache Solr 搜索平台，包括 spring-data-solr。

- spring-boot-starter-freemarker：支持 FreeMarker 模板引擎。
- spring-boot-starter-groovy-templates：支持 Groovy 模板引擎。
- spring-boot-starter-hateoas：通过 spring-hateoas 支持基于 HATEOAS 的 RESTful Web 服务。
- spring-boot-starter-integration：支持通用的 spring-integration 模块。
- spring-boot-starter-jdbc：支持 JDBC 数据库。
- spring-boot-starter-jersey：支持 Jersey RESTful Web 服务框架。
- spring-boot-starter-hornetq：通过 HornetQ 支持 JMS。
- spring-boot-starter-jta-atomikos：通过 Atomikos 支持 JTA 分布式事务处理。
- spring-boot-starter-jta-bitronix：通过 Bitronix 支持 JTA 分布式事务处理。
- spring-boot-starter-mail：支持 javax.mail 模块。
- spring-boot-starter-mobile：支持 spring-mobile。
- spring-boot-starter-mustache：支持 Mustache 模板引擎。
- spring-boot-starter-security：支持 spring-security。
- spring-boot-starter-social-facebook：支持 spring-social-facebook。
- spring-boot-starter-social-linkedin：支持 spring-social-linkedin。
- spring-boot-starter-social-twitter：支持 spring-social-twitter。
- spring-boot-starter-test：支持常规的测试依赖，包括 JUnit、Hamcrest、Mockito 以及 spring-test 模块。
- spring-boot-starter-thymeleaf：支持 Thymeleaf 模板引擎，包括与 Spring 的集成。
- spring-boot-starter-velocity：支持 Velocity 模板引擎。
- spring-boot-starter-web：支持全栈式 Web 开发，包括 Tomcat 和 spring-webmvc。
- spring-boot-starter-websocket：支持 WebSocket 开发。
- spring-boot-starter-ws：支持 Spring Web Services。

2.3 Spring Boot 自动配置原理

Spring Boot 在进行 SpringApplication 对象实例化时会加载 META-INF/spring.factories 文件，将该配置文件中的配置载入 Spring 容器，进行自动配置。

2.3.1 源码分析

首先进入启动 Spring Boot 项目代码 SpringApplication.run(App.class, args)的源码。

程序清单：org/springframework/boot/SpringApplication.java

```
public static ConfigurableApplicationContext run(Object[] sources, String[] args) {
    return new SpringApplication(sources).run(args);
}
```

可以看到 run 方法实际上在创建 SpringApplication 对象实例，下面来看创建 SpringApplication 对象实例的代码。

程序清单：org/springframework/boot/SpringApplication.java

```
public SpringApplication(Object... sources) {
    initialize(sources);
}
```

接下来就是调用 initialize(sources)方法，该方法的源码如下。

程序清单：org/springframework/boot/SpringApplication.java
```java
@SuppressWarnings({ "unchecked", "rawtypes" })
private void initialize(Object[] sources) {
    if (sources != null && sources.length > 0) {
        this.sources.addAll(Arrays.asList(sources));
    }
    this.webEnvironment = deduceWebEnvironment();
    setInitializers((Collection) getSpringFactoriesInstances(
            ApplicationContextInitializer.class));
    setListeners((Collection) getSpringFactoriesInstances(ApplicationListener.class));
    this.mainApplicationClass = deduceMainApplicationClass();
}
```

initialize 方法中调用了 getSpringFactoriesInstances 方法，代码如下。

程序清单：org/springframework/boot/SpringApplication.java
```java
private <T> Collection<? extends T> getSpringFactoriesInstances(Class<T> type,
        Class<?>[] parameterTypes, Object... args) {
    ClassLoader classLoader = Thread.currentThread().getContextClassLoader();
    // Use names and ensure unique to protect against duplicates
    Set<String> names = new LinkedHashSet<String>(
            SpringFactoriesLoader.loadFactoryNames(type, classLoader));
    List<T> instances = createSpringFactoriesInstances(type, parameterTypes,
            classLoader, args, names);
    AnnotationAwareOrderComparator.sort(instances);
    return instances;
}
```

在 getSpringFactoriesInstances 中又调用了 loadFactoryNames 方法，继续进入该方法，查看源码。

程序清单：org/springframework/boot.SpringApplication.java
```java
public static List<String> loadFactoryNames(Class<?> factoryClass, ClassLoader classLoader) {
    String factoryClassName = factoryClass.getName();
    try {
        Enumeration<URL> urls = (classLoader != null ? classLoader.getResources(FACTORIES_RESOURCE_LOCATION) :
            ClassLoader.getSystemResources(FACTORIES_RESOURCE_LOCATION));
        List<String> result = new ArrayList<String>();
        while (urls.hasMoreElements()) {
            URL url = urls.nextElement();
            Properties properties = PropertiesLoaderUtils.loadProperties(new UrlResource(url));
            String factoryClassNames = properties.getProperty(factoryClassName);
            result.addAll(Arrays.asList(StringUtils.commaDelimitedListToStringArray(factoryClassNames)));
        }
        return result;
    }
    catch (IOException ex) {
        throw new IllegalArgumentException("Unable to load [" + factoryClass.getName()
            + "] factories from location [" + FACTORIES_RESOURCE_LOCATION + "]", ex);
    }
}
```

在上述源码中可以看到加载了一个常量：FACTORIES_RESOURCE_LOCATION，该常量的源码如下。

```
/**
```

```
 * The location to look for factories.
 * <p>Can be present in multiple JAR files.
 */
public static final String FACTORIES_RESOURCE_LOCATION = "META-INF/spring.factories";
```

从该源码中可以看出，最终 Spring Boot 是通过加载 META-INF/spring.factories 文件进行自动配置的。其所在位置如图 2.9 所示。

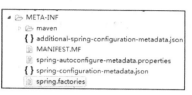

图 2.9　spring.factories 位置

▶▶ 2.3.2　spring.factories 分析

spring factories 文件非常重要，用来指导 Spring Boot 找到指定的自动配置文件。

spring factories 文件重点内容分析如下。

```
# PropertySource Loaders
org.springframework.boot.env.PropertySourceLoader=\
org.springframework.boot.env.PropertiesPropertySourceLoader,\
org.springframework.boot.env.YamlPropertySourceLoader
```

org.springframework.boot.env.PropertySourceLoader 对应的值表示指定 Spring Boot 配置文件支持的格式。Spring Boot 的配置文件内置支持 properties、xml、yml 和 yaml 几种格式。其中 properties 和 xml 对应的 Loader 类为 PropertiesPropertySourceLoader，yml 和 yaml 对应的 Loader 类为 YamlPropertySourceLoader。

```
# Run Listeners
org.springframework.boot.SpringApplicationRunListener=\
org.springframework.boot.context.event.EventPublishingRunListener
```

org.springframework.boot.SpringApplicationRunListener 对应的值表示运行的监听器类。默认会加载 EventPublishingRunListener，这个 RunListener 是在 SpringApplication 对象的 run 方法执行到不同的阶段时，发布相应的 event 给 SpringApplication 对象的 Listeners 中记录的事件监听器。

```
# Application Context Initializers
org.springframework.context.ApplicationContextInitializer=\
org.springframework.boot.context.ConfigurationWarningsApplicationContextInitializer,\
org.springframework.boot.context.ContextIdApplicationContextInitializer,\
org.springframework.boot.context.config.DelegatingApplicationContextInitializer,\
org.springframework.boot.context.embedded.ServerPortInfoApplicationContextInitializer
```

org.springframework.context.ApplicationContextInitializer 对应的值表示 Spring Boot 中的应用程序的初始化类。默认会加载 4 个 ApplicationContextInitializer 类。

- ➢ ConfigurationWarningsApplicationContextInitializer 的作用是报告常见的配置错误。
- ➢ ContextIdApplicationContextInitializer 的作用是给 ApplicationContext 设置一个 ID。
- ➢ DelegatingApplicationContextInitializer 的作用是将初始化的工作委托给 context.initializer.classes 环境变量指定的初始化器。
- ➢ ServerPortInfoApplicationContextInitializer 的作用是监听 EmbeddedServletContainer-InitializedEvent 类型的事件，然后将内嵌的 Web 服务器使用的端口设置到 Application-Context 中。

```
# Application Listeners
org.springframework.context.ApplicationListener=\
org.springframework.boot.ClearCachesApplicationListener,\
org.springframework.boot.builder.ParentContextCloserApplicationListener,\
org.springframework.boot.context.FileEncodingApplicationListener,\
org.springframework.boot.context.config.AnsiOutputApplicationListener,\
org.springframework.boot.context.config.ConfigFileApplicationListener,\
org.springframework.boot.context.config.DelegatingApplicationListener,\
org.springframework.boot.liquibase.LiquibaseServiceLocatorApplicationListener,\
org.springframework.boot.logging.ClasspathLoggingApplicationListener,\
org.springframework.boot.logging.LoggingApplicationListener
```

org.springframework.context.ApplicationListener 对应的值表示 Spring Boot 中的应用程序监听器，共有 9 个。

- org.springframework.boot.ClearCachesApplicationListener 的作用是在 Spring 的 context 容器完成 refresh()方法时调用，用来清除缓存信息。
- ParentContextCloserApplicationListener 的作用是当容器关闭时发出通知，如果父容器关闭，那么子容器也一起关闭。
- FileEncodingApplicationListener 的作用是在 Spring Boot 环境准备完成以后运行，获取环境中的系统环境参数，检测当前系统环境的 file.encoding 和 spring.mandatory-file-encoding 设置的值是否一样，如果不一样则抛出异常。
- AnsiOutputApplicationListener 的作用是在 Spring Boot 环境准备完成以后运行，如果终端支持 ANSI，设置的彩色输出会让日志更具可读性。
- ConfigFileApplicationListener 的作用是读取加载 Spring Boot 的配置文件如 application.properties 等，非常重要。
- DelegatingApplicationListener 的作用是把 Listener 转发给配置的 class 处理，这样可以支持外围代码不去改写 spring.factories 中的 org.springframework.context.ApplicationListener 的相关配置，保持 Spring Boot 代码的稳定性。
- LiquibaseServiceLocatorApplicationListener 的作用不大，指的是如果相关的参数 liquibase.servicelocator.ServiceLocator 存在，则使用 Spring Boot 相关的版本进行替代。
- ClasspathLoggingApplicationListener 的作用是程序启动时，将 classpath 打印到 debug 日志，启动失败时将 classpath 打印到 info 日志。
- LoggingApplicationListener 的作用是根据配置初始化日志系统进行日志输出。

```
# Environment Post Processors
org.springframework.boot.env.EnvironmentPostProcessor=\
org.springframework.boot.cloud.CloudFoundryVcapEnvironmentPostProcessor,\
org.springframework.boot.env.SpringApplicationJsonEnvironmentPostProcessor
```

org.springframework.boot.env.EnvironmentPostProcessor 对应的值表示 Spring Boot 中支持的动态读取文件方式，默认有两种方式。

- CloudFoundryVcapEnvironmentPostProcessor 的作用是对 CloudFoundry 提供支持。
- SpringApplicationJsonEnvironmentPostProcessor 的作用是把环境中 spring.application.json 的 json 值转化为 MapPropertySource，并将这个 MapPropertySource 添加到环境的属性源列表中。

通过源代码分析可以看出，Spring Boot 是通过 SpringFactoriesLoader 的 loadFactoryNames 方法读取 spring.factories 文件的。SpringFactoriesLoader 是 Spring 框架的一个工具类，属于 Spring 框架私有的一种扩展方案，它的主要功能就是从指定的配置文件 META-INF/spring.factories 加载配置。配合@EnableAutoConfiguration 使用的话，它更多是提供一种配置查找的功能支持，即根据@EnableAutoConfiguration 的完整类名 org.springframework.boot.autoconfigure.EnableAutoConfiguration 作为查找的 Key，获取对应的一组@Configuration 类。

所以，@SpringBootApplication 注解通过使用@EnableAutoConfiguration 注解自动配置的原理其实很简单：从 classpath 中搜寻所有的 META-INF/spring.factories 配置文件，并将其中 org.springframework.boot.autoconfigure.EnableAutoConfiguration 对应的配置项通过反射（Java Reflection）实例化为对应的标注了@Configuration 的 IoC 容器配置类，然后汇总并加载到 Spring 框架的 IoC 容器。

▶▶ 2.3.3 Spring Boot Web 开发的自动配置

在 spring.factories 中可以看出，Web 开发的自动配置类是 org.springframework.boot.autoconfigure.web.WebMvcAutoConfiguration，在这个类中自动实现了 Spring MVC 的配置。现在以 Spring MVC 的如下配置为例，了解 Spring Boot 是如何实现该自动配置的。

```xml
<bean id="jspViewResolver" class="org.springframework.web.servlet.view.InternalResourceViewResolver">
    <property name="prefix" value="/WEB-INF/jsp/"/>
    <property name="suffix" value=".jsp"/>
</bean>
```

① 查询 WebMvcAutoConfiguration 的源码如下。

```java
@Configuration
@ConditionalOnWebApplication
@ConditionalOnClass({ Servlet.class, DispatcherServlet.class,
    WebMvcConfigurerAdapter.class })
@ConditionalOnMissingBean(WebMvcConfigurationSupport.class)
@AutoConfigureOrder(Ordered.HIGHEST_PRECEDENCE + 10)
@AutoConfigureAfter({ DispatcherServletAutoConfiguration.class,
    ValidationAutoConfiguration.class })
public class WebMvcAutoConfiguration {
```

其中@ConditionalOnClass 是一个条件注解，意思就是只有当前项目运行环境中有 Servlet 类，并且有 DispatcherServlet 类以及 WebMvcConfigurerAdapter 类（说明本项目是需要集成 Spring MVC 的），Spring Boot 才会初始化 WebMvcAutoConfiguration 进行自动配置。

② 自动配置视图解析器 ViewResolver。

在 WebMvcAutoConfiguration 类下找到以下源码：

```java
@Bean
@ConditionalOnMissingBean
public InternalResourceViewResolver defaultViewResolver() {
    InternalResourceViewResolver resolver = new InternalResourceViewResolver();
    resolver.setPrefix(this.mvcProperties.getView().getPrefix());
    resolver.setSuffix(this.mvcProperties.getView().getSuffix());
    return resolver;
}
```

@Bean 在这里定义了一个 Bean 对象 InternalResourceViewResolver，以前是通过<bean/>标签来定义的。

@ConditionalOnMissingBean 是一个条件注解，在当前环境下没有这个 Bean 的时候才会创建该 Bean。

方法的返回值即 InternalResourceViewResolver，正是我们需要的对象，那么视图解析器中前缀和后缀 Spring Boot 是如何实现自动配置的呢？

```java
resolver.setPrefix(this.mvcProperties.getView().getPrefix());
resolver.setSuffix(this.mvcProperties.getView().getSuffix());
```

该源码会找到一个 View 对象：

```java
public static class View {
    /**
     * Spring MVC 视图前缀
     */
    private String prefix;

    /**
     * Spring MVC 视图后缀
     */
    private String suffix;

    public String getPrefix() {
        return this.prefix;
    }

    public void setPrefix(String prefix) {
        this.prefix = prefix;
    }

    public String getSuffix() {
        return this.suffix;
    }

    public void setSuffix(String suffix) {
        this.suffix = suffix;
    }
}
```

View 对象则会通过获取 prefix、suffix 加载视图解析器需要的前缀和后缀，该参数的值是可以通过全局配置文件来指定前缀和后缀的，配置如下：

```
spring.mvc.view.prefix= # Spring MVC view prefix
spring.mvc.view.suffix= # Spring MVC view suffix
```

2.4　本章小结

本章主要介绍了 Spring Boot 的核心注解、基本配置和自动配置的原理和运行机制。精通一项技术一定要深入了解这项技术帮助我们做了哪些工作，深入理解它的底层运行原理，只有达到这个目标才可以熟练使用框架，最终才能融会贯通。

第 3 章将重点介绍 Spring Boot 的 Web 开发。

CHAPTER 3

第 3 章
Spring Boot 的 Web 开发

本章要点

- Spring Boot 的 Web 开发支持
- Thymeleaf 模板引擎介绍
- Spring 和 Thymeleaf 的整合
- Spring Boot 对 Thymeleaf 支持
- Spring Boot 对 JSP 的支持
- Spring Boot 对 JSON 的支持
- Spring Boot 的文件上传下载
- Spring Boot 的异常处理

Web开发是现代软件开发中最重要的一部分,现在几乎所有的商业软件都要基于Web开发。Spring Boot的Web开发内嵌了Servlet容器,并结合Spring MVC来完成。

3.1 Spring Boot的Web开发支持

Spring Boot使用spring-boot-starter-web为Web开发提供支持,spring-boot-starter-web包含了Spring Boot预定义的一些Web开发的常用依赖包,为开发者提供了嵌入的Tomcat以及Spring MVC的依赖。而与Web相关的自动配置支持则保存在spring-boot-autoconfigure.jar的org.springframework.boot.autoconfigure.web下面,如图3.1所示。

图3.1 Web相关的自动配置支持类

重要的自动配置支持如下:
➢ HttpEncodingAutoConfiguration和HttpEncodingProperties用来自动配置http的编码。
➢ JacksonHttpMessageConvertersConfiguration用来自动配置Jackson相关的Converter。
➢ MultipartAutoConfiguration和MultipartProperties用来自动配置上传文件的属性。
➢ ServerPropertiesAutoConfiguration和ServerProperties用来自动配置内嵌的Servlet容器。
➢ WebMvcAutoConfiguration和WebMvcProperties用来自动配置Spring MVC功能。

3.2 Thymeleaf模板引擎

Spring Boot建议使用HTML来完成动态页面。Spring Boot提供了大量的模板引擎,包括Thymeleaf、FreeMarker、Velocity等。

Spring Boot官方推荐使用Thymeleaf模板引擎来完成动态页面,并且为Thymeleaf提供了完美的Spring MVC的支持,Thymeleaf模板引擎可以支持纯HTML浏览器展现(模板表达式在脱离运行环境下不污染HTML结构)。

3.2.1 Thymeleaf 概述

Thymeleaf 是面向 Web 和独立环境的现代服务器端 Java 模板引擎，能够处理 HTML、XML、JavaScript、CSS 甚至纯文本，可以作为 MVC Web 应用层的 View 层显示数据。

Thymeleaf 是一个扩展性很强的模板引擎（实际上它可以称为模板引擎框架），Thymeleaf 允许开发者自定义模板，并且很好地处理该模板的细节。将一些逻辑应用于标记组件（标签、某些文本、注释或只有占位符）的一个对象被称为处理器，通常这些处理器的集合以及一些额外的组件就组成了 Thymeleaf 方言。Thymeleaf 的核心库提供了一种称为标准方言的方言，这对大多数用户来说已经足够了。如果用户希望在使用标准方言库的高级功能的同时定义自己的处理逻辑，也可以创建自己的方言（甚至扩展标准的方言）。

官方的 thymeleaf-spring5 的整合包里定义了一种称为"Spring 标准方言"的方言，该方言与"Thymeleaf 标准方言"大致相同，但是对于 Spring 框架中的某些功能（例如，通过使用 SpringEL 表达式代替 OGNL 表达式）做了一些简单的调整。使之更完美地匹配 Spring MVC，所以 Spring Boot 官方推荐使用 Thymeleaf 完全替代 JSP。

Thymeleaf 标准方言中的大多数处理器都是属性处理器。这样，即使在模板未被处理之前，浏览器也可以正确地显示 HTML 模板文件，因为浏览器将简单地忽略其不识别的属性。例如，像下面这段 JSP 模板的代码片段就不能在模板被解析之前通过浏览器直接显示。

```
<form:inputText name="username" value="${user.username}" />
```

然而 Thymeleaf 标准方言将允许实现与上述代码相同的功能。

```
<input type="text" name="username" value="fkjava" th:value="${user.username}" />
```

浏览器不仅可以正确显示这些信息，而且还可以在浏览器中静态打开时显示一个默认的值（在本例中为"fkjava"），在模板处理期间由 ${user.username} 的值代替 value 的默认值。这样有助于设计师和开发人员处理相同的模板文件，并减少将静态原型转换为工作模板文件所需的工作量。具备这种能力的模板被称为自然模板。

3.2.2 Thymeleaf 基础语法

3.2.2.1 引入 Thymeleaf

首先要像如下所示改写 html 标签。

```
<html xmlns:th="http://www.thymeleaf.org">
```

这样才可以在其他标签里面使用 th:*这样的语法。

通过 xmlns:th="http://www.thymeleaf.org"命名空间，引入 Thymeleaf 模板引擎，将静态页面转换为动态页面。需要进行动态处理的元素都使用"th:"为前缀。

3.2.2.2 引入 URL

Thymeleaf 对于 URL 的处理是通过语法@{…}来处理的。

```
<a th:href="@{http://www.fkit.org/fkjava.png}">绝对路径</a>
<a th:href="@{/}">相对路径</a>
<a th:href="@{css/bootstrap.min.css}">Content 路径，默认访问 static 下的 css 文件夹</a>
```

3.2.2.3 表达式

有一些专门的表达式，用来从模板中的 WebContext 获取请求参数、请求、会话和应用程序中的属性，这些操作和 JSP 的 EL 表达式非常相似。

${x}将返回存储在 Thymeleaf 上下文中的变量 x 或作为请求 Request 作用域中的属性。

$ {param.x}将返回一个名为 x 的请求参数（可能是多值的）。
$ {session.x}将返回一个名为 x 的会话 HttpSession 作用域中的属性。
$ {application.x}将返回一个名为 x 的全局 ServletContext 上下文作用域中的属性。

获取变量值用$符号，对于 JavaBean 使用"变量名.属性名"方式获取，这点和 EL 表达式一样。另外，$表达式只能写在 th 标签内部，否则不会生效。

3.2.2.4 字符串操作

很多时候可能只需对一大段文字中的某一处进行替换，可以通过字符串拼接操作完成。

```
<span th:text="'Welcome to fkit, ' + ${user.name} + '!'">
```

还有一种更简捷的方式。

```
<span th:text="|Welcome to fkit, ${user.name}!|">
```

这种形式限制比较多，|…|中只能包含变量表达式${…}，不能包含其他常量、条件表达式等。

3.2.2.5 运算符

在表达式中可以使用各类算术运算符，例如+、-、*、/、%。

```
th:with="isEven=(${prodStat.count} % 2 == 0)"
```

逻辑运算符>、<、<=、>=、==、!=都可以使用，唯一需要注意的是，使用<>时应用它的 HTML 转义符。

```
th:if="${prodStat.count} &gt; 1"
th:text="'Execution mode is ' + ((${execMode} == 'dev')? 'Development' : 'Production')"
```

3.2.2.6 条件判断

Thymeleaf 中使用 th:if 和 th:unless 属性进行条件判断，标签只有在 th:if 中的条件成立时才显示，th:unless 与 th:if 恰好相反，只有表达式中的条件不成立时，才会显示其内容。

```
<a th:href="main.html" th:if="${username != null}">Login1</a><br>
<a th:href="main.html" th:unless="${username != null}">Login2</a><br>
```

Thymeleaf 同样支持多路选择 Switch 结构，默认属性 default 可以用*表示。

```
<div th:switch="${role}">
  <p th:case="'admin'">User is an administrator</p>
  <p th:case="'manager'">User is a manager</p>
  <p th:case="*">User is some other thing</p>
</div>
```

3.2.2.7 循环

Thymeleaf 中循环变量集合使用 th:each 标签。

th:each 属性用于迭代循环，迭代对象可以是 java.util.List、java.util.Map 或数组等，语法如下。

```
th:each="obj,iterStat:${objList}"
```

下面是一个简单循环。

```
<tr th:each="book : ${books}">
    <td th:text="${book.title}">疯狂 Spring Boot 讲义</td>
    <td th:text="${boot.author}">疯狂软件编著</td>
    <td th:text="${boot.remark}">最新的 Spring Boot 书籍!</td>
</tr>
```

iterStat 称作状态变量，有以下这些属性。

- index：当前迭代对象的 index（从 0 开始计算）。
- count：当前迭代对象的 index（从 1 开始计算）。
- size：被迭代对象的大小。
- current：当前迭代变量。
- even/odd：布尔值，当前循环是否是偶数/奇数（从 0 开始计算）。
- first：布尔值，当前循环是否是第一个。
- last：布尔值，当前循环是否是最后一个。

```html
<table border="1">
    <tr>
    <th>用户名</th>
    <th>邮箱</th>
    <th>状态变量：index</th>
    <th>状态变量：count</th>
    <th>状态变量：size</th>
    <th>状态变量：current.username</th>
    <th>状态变量：even</th>
    <th>状态变量：odd</th>
    <th>状态变量：first</th>
    <th>状态变量：last</th>
    </tr>
    <tr th:each="user,userStat : ${list}">
    <td th:text="${user.username}">admin</td>
    <td th:text="${user.email}">fkit@qq.com.cn</td>
    <th th:text="${userStat.index}">状态变量：index</th>
    <th th:text="${userStat.count}">状态变量：count</th>
    <th th:text="${userStat.size}">状态变量：size</th>
    <th th:text="${userStat.current.username}">状态变量：current</th>
    <th th:text="${userStat.even}">状态变量：even****</th>
    <th th:text="${userStat.odd}">状态变量：odd</th>
    <th th:text="${userStat.first}">状态变量：first</th>
    <th th:text="${userStat.last}">状态变量：last</th>
    </tr>
</table>
```

3.2.2.8 内置对象

Web 项目中经常会传递一些数据，例如 String 类型的列表、日期对象和数值，这些在实际应用开发中应用非常广泛，所以 Thymeleaf 还提供了很多内置对象，可以通过#直接访问。

> **注意**
> 内置对象一般都以 s 结尾，如 dates、lists、numbers 等。在使用内置对象时需要在对象名前加#号。

在 Thymeleaf 中的内置对象有以下这些。
- #dates：日期格式化内置对象，具体方法可以参照 java.util.Date。
- #calendars：类似于#dates，但是是 java.util.Calendar 类的方法。
- #numbers：数字格式化。
- #strings：字符串格式化，具体方法可以参照 java.lang.String，如 startsWith、contains 等。
- #objects：参照 java.lang.Object。
- #bools：判断 boolean 类型的工具。

- #arrays：数组操作的工具。
- #lists：列表操作的工具，参照 java.util.List。
- #sets：Set 操作工具，参照 java.util.Set。
- #maps：Map 操作工具，参照 java.util.Map。
- #aggregates：操作数组或集合的工具。
- #messages：操作消息的工具。

示例代码如下。

```
<span th:text="${#dates.format(curDate, 'yyyy-MM-dd HH:mm:ss')}"></span>
```

使用内置对象 dates 的 format 函数即可对日期进行格式化，在 format 函数中，第一个参数是日期对象，第二个参数为日期格式（规则与 SimpleDateFormat 一样）。

```
<span th:text="${#numbers.formatDecimal(money, 0, 2)}"></span>
```

以上代码表示保留 2 位小数，整数位自动。

```
<span th:text="${#numbers.formatDecimal(money, 3, 2)}"></span>
```

以上代码表示保留 2 位小数，3 位整数（不够的在前面加 0）。

```
<span th:text="${#strings.length(str)}"></span>
```

以上代码表示获取变量 str 的长度。

```
<span th:text="${#lists.size(datas)}"></span>
```

以上代码表示使用#lists.size 来获取名称为 datas 的集合的长度。

3.3　Spring 和 Thymeleaf 的整合

Thymeleaf 支持和 Spring 框架的集成，最新版本的支持封装在 thymeleaf-spring5 这个独立的库中。

Thymeleaf 与 Spring 进行整合后，可以在 Spring MVC 应用中完全替代 JSP 文件。

- 就像控制 JSP 一样，使用 Spring MVC 的@Controller 注解来映射 Thymeleaf 的模板文件。
- 在模板中使用 SpringEL 表达式来替换 OGNL。
- 在模板中创建的表单，完全支持 Beans 和结果的绑定，包括使用 PropertyEditor、转换和验证等。
- 可以通过 Spring 来管理国际化文件显示国际化信息。

为了更加方便、快捷地集成，Thymeleaf 提供了一套能够与 Spring 正确工作的特有方言。这套方言基于 Thymeleaf 标准方言实现，它在类 org.thymeleaf.spring.dialect.SpringStandardDialect 中，事实上，它继承自 org.thymeleaf.standard.StandardDialect。除了已经出现在 Thymeleaf 标准方言中的所有功能，Thymeleaf 的 Spring 方言中还有以下特点：

- Thymeleaf 不适用 OGNL，而是 SpringEL 实现变量表达式，因此，所有的${...}和*{...}表达式将用 Spring 的表达式引擎进行处理。
- 访问应用 context 中的 Beans 可以使用 SpringEL 语法：${@myBean.doSomething()}。
- 基于表格处理的新属性：th:field、th:errors 和 th:errorclass，此外还有一个 th:object 的新实现，允许它使用表单命令选择器。
- 在 Spring 5.0 集成中提供多个新的表达式。

Thymeleaf 提供了一个 org.thymeleaf.spring4.SpringTemplateEngine 类，用来执行在 Spring MVC 中使用 Thymeleaf 模板引擎，另外还提供了一个 TemplateResolver 用来设置通用的模板引擎，如前缀、后缀等，这使得开发者在 Spring MVC 中集成 Thymeleaf 变得非常简单。

3.4 Spring Boot 的 Thymeleaf 支持

上一节介绍了 Spring 和 Thymeleaf 的整合，但是这些工作在 Spring Boot 中都不再需要开发者自行配置，Spring Boot 通过 org.springframework.boot.autoconfigure.thymeleaf 包对 Thymeleaf 进行自动配置，如图 3.2 所示。

图 3.2 Thymeleaf 包下的具体类

Thymeleaf 包下最重要的类是 ThymeleafAutoConfiguration 和 ThymeleafProperties。

ThymeleafAutoConfiguration 类对集成所需要的 Bean 进行自动配置，包括 templateEngine 和 templateResolver 的配置。

ThymeleafProperties 类读取 application.properties 配置文件，设置 Thymeleaf 的属性以及默认配置。ThymeleafProperties 类的重点源代码如下。

程序清单：org/springframework/boot/autoconfigure/thymeleaf/ThymeleafProperties

```java
@ConfigurationProperties(prefix = "spring.thymeleaf")
public class ThymeleafProperties {

    private static final Charset DEFAULT_ENCODING = Charset.forName("UTF-8");

    public static final String DEFAULT_PREFIX = "classpath:/templates/";

    public static final String DEFAULT_SUFFIX = ".html";

    /**
     * 检查 templates 路径是否存在
     */
    private boolean checkTemplateLocation = true;

    /**
     * 前缀设置，引用上面的常量 DEFAULT_PREFIX
     * 即默认放置在 classpath:/templates/ 目录下
     */
    private String prefix = DEFAULT_PREFIX;

    /**
     * 后缀设置，引用上面的常量 DEFAULT_SUFFIX
     * 即默认为 html
     */
    private String suffix = DEFAULT_SUFFIX;

    /**
     * 模板模式设置，默认为 HTML
     */
    private String mode = "HTML";

    /**
```

```
     * 模板的编码设置,引用上面的常量 DEFAULT_ENCODING
     * 即默认为 UTF-8
     */
    private Charset encoding = DEFAULT_ENCODING;

    /**
     * 模板的类型设置,默认为 text/html
     */
    private String contentType = "text/html";

    /**
     * 是否开启模板缓存,默认开启,开发时通常关闭
     */
    private boolean cache = true;

    // ...
}
```

ThymeleafProperties 类中设置了 Thymeleaf 的默认设置,项目中可以通过 application.properties 配置文件对默认设置进行修改。

```
#thymeleaf start
spring.thymeleaf.prefix=classpath:/templates/
spring.thymeleaf.suffix=.html
spring.thymeleaf.mode=HTML5
spring.thymeleaf.encoding=UTF-8
spring.thymeleaf.content-type=text/html
#开发时关闭缓存,不然没法看到实时页面
spring.thymeleaf.cache=false
#thymeleaf end
```

如果提供了 application.properties 配置文件,ThymeleafProperties 会读取 application.properties 配置文件中的配置信息作为 Thymeleaf 的设置属性。

3.5 Spring Boot 的 Web 开发实例

本书的 Web 项目开发中大量使用了前端开发的开源工具包 Bootstrap、优秀的 JavaScript 框架 jQuery 和流行的 MVC 框架 Spring MVC。

关于 Bootstrap 和 jQuery 的具体用法,请参考疯狂软件系列图书之《疯狂前端开发讲义》。

关于 Spring MVC 的具体用法,请参考疯狂软件系列之《Spring+MyBatis 企业应用实战》。

示例:第一个 Spring Boot 的 Web 应用

创建一个新的 Maven 项目,命名为 logintest。按照 Maven 项目的规范,在 src/main/下新建一个名为 resources 的文件夹,并在 src/main/resources 下再新建 static 和 templates 两个文件夹。

① 修改 pom.xml 文件。

程序清单:codes/03/logintest/pom.xml

```
<project xmlns="http://maven.apache.org/POM/4.0.0" xmlns:xsi="http://www.w3.org/2001/XMLSchema-instance"
    xsi:schemaLocation="http://maven.apache.org/POM/4.0.0 http://maven.apache.org/xsd/maven-4.0.0.xsd">
    <modelVersion>4.0.0</modelVersion>
```

```xml
    <groupId>org.fkit.springboot</groupId>
    <artifactId>logintest</artifactId>
    <version>0.0.1-SNAPSHOT</version>
    <packaging>jar</packaging>

    <name>logintest</name>
    <url>http://maven.apache.org</url>

    <parent>
        <groupId>org.springframework.boot</groupId>
        <artifactId>spring-boot-starter-parent</artifactId>
        <version>2.0.0.RELEASE</version>
    </parent>

<properties>
    <project.build.sourceEncoding>UTF-8</project.build.sourceEncoding>
    <project.reporting.outputEncoding>UTF-8</project.reporting.outputEncoding>
    <java.version>1.8</java.version>
</properties>
    <dependencies>
        <!-- 添加 spring-boot-starter-web 模块依赖 -->
        <dependency>
            <groupId>org.springframework.boot</groupId>
            <artifactId>spring-boot-starter-web</artifactId>
        </dependency>

        <!-- 添加 spring-boot-starter-thymeleaf 模块依赖 -->
        <dependency>

        <dependency>
            <groupId>org.springframework.boot</groupId>
            <artifactId>spring-boot-starter-thymeleaf</artifactId>
        </dependency>

        <dependency>
          <groupId>junit</groupId>
          <artifactId>junit</artifactId>
          <scope>test</scope>
        </dependency>
    </dependencies>

</project>
```

pom.xml 文件引入 thymeleaf 依赖。

```xml
<dependency>
    <groupId>org.springframework.boot</groupId>
    <artifactId>spring-boot-starter-thymeleaf</artifactId>
</dependency>
```

> **注意**
> 在 Spring Boot 1.x 版本当中，一些功能模块 Starter 之间是存在包含引用关系的，spring-boot-starter-thymeleaf 中就包含了 spring-boot-starter-web 模块。但是，在 Spring 5 中，WebFlux 的出现对于 Web 应用的解决方案将不再唯一，因此 spring-boot-starter-thymeleaf 中的依赖就不再包含 spring-boot-starter-web，开发人员需要自己添加 spring-boot-starter-web 模块依赖。

② 引入静态文件。

将 Bootstrap 的脚本样式、项目所需的图片等静态文件放在 src/main/resources/static 下，在 src/main/resources/templates 下新建一个页面文件 index.html，如图 3.3 所示。

图 3.3　Web 项目结构

③ 生成静态页面。

程序清单：codes/03/logintest/src/main/resources/templates/index.html

```html
<!DOCTYPE html>
<html xmlns:th="http://www.thymeleaf.org">
<head>
<meta charset="UTF-8"></meta>
<title>Insert title here</title>
<link rel="stylesheet" th:href="@{css/bootstrap.min.css}" />
<link rel="stylesheet" th:href="@{css/bootstrap-theme.min.css}"/>
<script type="text/javascript" th:src="@{js/jQuery-1.11.0.min.js}"></script>
<script type="text/javascript" th:src="@{js/bootstrap.min.js}"></script>
<script type="text/javascript">
    $(function(){
        $("#loginbtn").click(function(){
            var loginName = $("#loginName");
            var password = $("#password");
            var msg = "";
            if(loginName.val() == ""){
                msg = "登录名不能为空!";
                loginName.focus();
            }else if(password.val() == ""){
                msg = "密码不能为空!";
                password.focus();
            }
            if(msg != ""){
                alert(msg);
            }else{
                $("#loginform").submit();
            }
        })
    })
</script>
</head>
<body>
<div class="panel panel-primary">
    <!-- .panel-heading 面板头信息。 -->
    <div class="panel-heading">
        <!-- .panel-title 面板标题。 -->
        <h3 class="panel-title">第一个 Spring Boot Web 开发示例</h3>
    </div>
```

```html
            </div>
            <!-- .container 类用于固定宽度并支持响应式布局的容器。 -->
            <div class="container">
                <!-- 栅格系统用于通过一系列的行（row）与列（column）的组合来创建页面布局 -->
                <div class="row">
                    <!-- 页面标题 -->
                    <div class="page-header">
                        <h2>用户登录</h2>
                        <!--
                        .form-horizontal:将 label 标签和控件组水平并排布局。
                        .form-inline:多个控件可以排列在同一行，水平
                        -->
                        <form class="form-horizontal" action="login" method="post" id="loginform">
                            <div class="form-group">
                                <!-- input-group用于将图片和控件放在一组 -->
                                <div class="input-group col-md-4">
                                    <!-- 额外的内容（图片）放在 input-group-addon 中-->
                                    <span class="input-group-addon"><i class="glyphicon glyphicon-user"></i></span>
                                    <input class="form-control" placeholder="用户名/邮箱" type="text" name="loginName" id="loginName"/>
                                </div>
                            </div>
                            <div class="form-group">
                                <div class="input-group col-md-4">
                                    <span class="input-group-addon"><i class="glyphicon glyphicon-lock"></i></span>
                                    <input class="form-control" placeholder="密码" type="password" name="password" id="password"/>
                                </div>
                            </div>
                            <div class="form-group">
                                <div class="col-md-4">
                                    <!--btn-group-justified 能够让按钮组根据父容器尺寸来设定各自相同的尺寸，采用响应式布局技术从而有利于移动端的用户体验。-->
                                    <div class="btn-group btn-group-justified" >
                                        <div class="btn-group" >
                                            <button type="button" class="btn btn-success" id="loginbtn">
                                                <span class="glyphicon glyphicon-log-in"></span>
                                                 登录</button>
                                        </div>
                                        <div class="btn-group" >
                                            <button type="button" class="btn btn-danger" id="registerbtn">
                                                <span class="glyphicon glyphicon-edit"></span>
                                                 注册</button>
                                        </div>
                                    </div>
                                </div>
                            </div>
                        </form>
                    </div>
                </div>
            </div>
    </body>
</html>
```

  index.html 中有两个输入框，分别输入登录名 loginName 和密码 password，当单击登录按钮时，使用 jQuery 完成输入验证，提交到名为 login 的请求。

程序清单：codes/03/logintest/src/main/resources/templates/main.html

```html
<!DOCTYPE html>
<html xmlns:th="http://www.thymeleaf.org">
<head>
<meta charset="UTF-8"></meta>
<link rel="stylesheet" th:href="@{css/bootstrap.min.css}" />
<link rel="stylesheet" th:href="@{css/bootstrap-theme.min.css}"/>
<script type="text/javascript" th:src="@{js/jQuery-1.11.0.min.js}"></script>
<script type="text/javascript" th:src="@{js/bootstrap.min.js}"></script>
<title>Insert title here</title>
</head>
<body>
<div class="panel panel-primary">
    <!-- .panel-heading 面板头信息。 -->
    <div class="panel-heading">
        <!-- .panel-title 面板标题。 -->
        <h3 class="panel-title">第一个 Spring Boot Web 开发示例</h3>
    </div>
</div>
    <div class="container">
        <div class="panel panel-primary">
            <!-- .panel-heading 面板头信息。 -->
            <div class="panel-heading">
                <!-- .panel-title 面板标题。 -->
                <h3 class="panel-title">图书信息列表</h3>
            </div>
            <div class="panel-body">
                <div class="container">
                    <div class="row">
                        <div class="col-md-4 col-sm-6">
                            <a href=""> <img src="img/java.jpg" alt="图书封面"/>
                            </a>
                            <div class="captio">
                                <h4>疯狂 Java 讲义</h4>
                                <p>李刚</p>
                                <p>电子工业出版社</p>
                                <p>109.00</p>
                            </div>
                        </div>

                        <div class="col-md-4 col-sm-6">
                            <a href=""> <img src="img/ee.jpg" alt="图书封面"/>
                            </a>
                            <div class="caption">
                                <h4>轻量级 Java EE 企业应用实战</h4>
                                <p>李刚 编著</p>
                                <p>电子工业出版社</p>
                                <p>108.00</p>
                            </div>
                        </div>

                        <div class="col-md-4 col-sm-6">
                            <a href=""> <img src="img/SpringMyBatis.jpg" alt="图书封面"/>
                            </a>
                            <div class="caption">
                                <h4>Spring+MyBatis 应用实战</h4>
                                <p>疯狂软件 编著</p>
                                <p>电子工业出版社</p>
                                <p>58.00</p>
                            </div>
                        </div>
```

```html
            <div class="col-md-4 col-sm-6">
                <a href=""> <img src="img/android.jpg" alt="图书封面"/>
                </a>
                <div class="caption">
                    <h4>疯狂 Android 讲义</h4>
                    <p>李刚 编著</p>
                    <p>电子工业出版社</p>
                    <p>108.00</p>
                </div>
            </div>
            <div class="col-md-4 col-sm-6">
                <a href=""> <img src="img/ajax.jpg" alt="图书封面"/>
                </a>
                <div class="caption">
                    <h4>疯狂 Ajax 开发</h4>
                    <p>李刚 编著</p>
                    <p>电子工业出版社</p>
                    <p>79.00</p>
                </div>
            </div>
        </div>
    </div>
  </div>
</body>
</html>
```

main.html 中使用 Bootstrap 框架完成页面排版，显示了疯狂软件系列的 5 本图书。

④ 处理请求的控制器类。

程序清单：codes/03/logintest/src/main/java/org/fkit/logintest/controller/IndexController.java

```java
import org.springframework.stereotype.Controller;
import org.springframework.ui.Model;
import org.springframework.web.bind.annotation.RequestMapping;

@Controller
public class IndexController {
    // 映射"/"请求
    @RequestMapping("/")
    public String index(Model model){
        System.out.println("IndexController index 方法被调用……");
        // 根据 Thymeleaf 默认模板，将返回 resources/templates/index.html
        return "index";
    }
}
```

@Controller 注解用于指示该类是一个控制器。@RequestMapping 注解用于映射请求的 URL。

返回的是字符串"index"，由于 Thymeleaf 默认的前缀是"classpath:/templates/"，后缀是"html"，所以该请求返回"classpath:/templates/index.html"。

程序清单：codes/03/logintest/src/main/java/org/fkit/logintest/controller/LoginController.java

```java
import org.springframework.stereotype.Controller;
import org.springframework.web.bind.annotation.PostMapping;
import org.springframework.web.bind.annotation.RequestParam;
import org.springframework.web.servlet.ModelAndView;
@Controller
```

```java
public class LoginController {
    @PostMapping("login")
    public ModelAndView login(
            @RequestParam("loginName") String loginName,
            @RequestParam("password") String password,
            ModelAndView mv){
        System.out.println("LoginController login 方法被调用……");
        System.out.println("LoginController 登录名:"+loginName + " 密码:" + password);
        // 重定向到 main 请求
        mv.setViewName("redirect:/main");
        return mv;
    }
}
```

LoginController 类接收 login 请求，重定向到 main 请求。

程序清单：codes/03/logintest/src/main/java/org/fkit/logintest/controller/MainController.java

```java
import org.springframework.stereotype.Controller;
import org.springframework.web.bind.annotation.RequestMapping;

@Controller
public class MainController {

    @RequestMapping(value="/main")
    public String main(){
        System.out.println("MainAction main 方法被调用……");
        // 根据 Thymeleaf 默认模板，将返回 resources/templates/main.html
        return "main";
    }

}
```

MainController 返回"classpath:/templates/main.html"。

① 测试应用。

App.java 和之前的项目一致，此处不再赘述。单击右键，从快捷菜单中选择"运行"命令运行 main 方法。

Spring Boot 项目启动后，在浏览器中输入 URL 来测试应用。

```
http://localhost:8080
```

请求会提交到 IndexController 类的 index 方法进行处理，该方法返回字符串"index"，即跳转到 templates/index.html 页面，如图 3.4 所示。

图 3.4 登录页面

输入任意登录名、密码，单击"登录"按钮。请求将会提交到 LoginController 类的 login 方法。该方法接收请求参数 loginName 和 password，并重定向到 main 请求。main 请求的处理类是 MainController 类，处理方法是 main，该方法返回字符串"main"，即跳转到 templates/main.html 页面，如图 3.5 所示。

图 3.5 网站首页

查看 Eclipse 控制台输出的结果，程序运行结果和之前描述的一致。

```
LoginController login 方法被调用……
LoginController 登录名:jack 密码:123456
MainController main 方法被调用……
```

本示例演示了 Spring Boot 的 Web 开发过程，包括 Spring MVC 的使用和 Thymeleaf 的集成，之后的项目都建立在这些基础之上。

示例：Thymeleaf 常用功能

创建一个新的 Maven 项目，命名为 thymeleaftest。按照 Maven 项目的规范，在 src/main/ 下新建一个名为 resources 的文件夹，并在 src/main/resources 下新建 static 和 templates 两个文件夹。

修改 pom.xml 文件，文件内容和之前示例一致，引入静态文件，操作和之前示例一致。

❶ 测试 Thymeleaf 表达式访问数据。

程序清单：codes/03/thymeleaftest/src/main/resources/templates/index.html

```html
<!DOCTYPE html>
<html xmlns:th="http://www.thymeleaf.org">
<head>
<meta charset="UTF-8"></meta>
<title>thymeleaf 示例</title>
<link rel="stylesheet" th:href="@{css/bootstrap.min.css}" />
<link rel="stylesheet" th:href="@{css/bootstrap-theme.min.css}"/>
<script type="text/javascript" th:src="@{js/jQuery-1.11.0.min.js}"></script>
<script type="text/javascript" th:src="@{js/bootstrap.min.js}"></script>
</head>
<body>
<div class="panel panel-primary">
    <!-- .panel-heading 面板头信息。 -->
    <div class="panel-heading">
        <!-- .panel-title 面板标题。 -->
        <h3 class="panel-title">thymeleaf 常用功能示例</h3>
    </div>
```

```html
        </div>
        <div class="container">
            <div class="row">
                <div class="col-md-4">
                    <a th:href="@{regexptest?loginName=jack&amp;password=123456}">测试表达
式访问数据</a><br/><br/>
                    <a th:href="@{iftest}">测试条件判断</a><br/><br/>
                    <a th:href="@{eachtest}">测试循环</a><br/><br/>
                </div>
            </div>
        </div>
    </body>
</html>
```

index.html 中有三个超链接，分别用来测试 thymeleaf 的表达式、条件判断和循环。

程序清单：codes/03/thymeleaftest/src/main/java/org/fkit/thymeleaftest/controller/IndexController.java

```java
import org.springframework.stereotype.Controller;
import org.springframework.web.bind.annotation.RequestMapping;
@Controller
public class IndexController {
    @RequestMapping("/")
    public String index(){
        System.out.println("IndexController index 方法被调用……");
        return "index";
    }
}
```

IndexController 类用来跳转到 index.html。

程序清单：codes/03/thymeleaftest/src/main/java/org/fkit thymeleaftest/controller/ThymeleafController.java

```java
import java.util.ArrayList;
import java.util.List;
import javax.servlet.http.HttpServletRequest;
import javax.servlet.http.HttpSession;
import org.fkit.thymeleaftest.domain.Book;
import org.springframework.stereotype.Controller;
import org.springframework.web.bind.annotation.RequestMapping;
import org.springframework.web.context.request.WebRequest;
import org.springframework.web.context.request.RequestAttributes;
@Controller
public class ThymeleafController {

    /*
     * 将数据保存到作用域，用于测试 Thymeleaf 表达式访问数据
     * */
    @RequestMapping("/regexptest")
    public String regexptest(
            HttpServletRequest request,
            HttpSession session) {
        // 将数据保存到 request 作用域当中
        request.setAttribute("book", "疯狂 Spring Boot 讲义");
        // 将数据保存到 session 作用域当中
        session.setAttribute("school", "疯狂软件");
        // 将数据保存到 ServletContext (application) 作用域当中
        request.getServletContext().setAttribute("name", "Thymeleaf 模板引擎");
        return "success1";
    }
}
```

ThymeleafController 类用来响应 index.html 中的请求，regexptest 方法用来响应第一个请求 `<a th:href="@{regexptest?loginName=jack&password=123456}">测试表达式访问数据</a>`，

regexptest 方法中设置了一个"book"变量到 request 作用域；设置了一个"school"变量到 session 作用域；设置了一个"name"变量到 ServletContext 全局作用域，然后返回 success1.html。

程序清单：codes/03/thymeleaftest/src/main/resources/templates/success1.html

```html
<!DOCTYPE html>
<html xmlns:th="http://www.thymeleaf.org">
<head>
<meta charset="UTF-8"></meta>
<title>thymeleaf 示例</title>
<link rel="stylesheet" th:href="@{css/bootstrap.min.css}" />
<link rel="stylesheet" th:href="@{css/bootstrap-theme.min.css}"/>
<script type="text/javascript" th:src="@{js/jQuery-1.11.0.min.js}"></script>
<script type="text/javascript" th:src="@{js/bootstrap.min.js}"></script>
</head>
<body>
<div class="panel panel-primary">
    <div class="panel-heading">
        <h3 class="panel-title">Thymeleaf 表达式访问数据</h3>
    </div>
</div>
<div class="container">
    <div class="row">
        <div class="col-md-4">
            <p><font color="red">${param.x}将返回一个名为 x 的请求参数</font></p><br/>
            访问页面传递的参数：<span th:text="${param.loginName[0]}">登录名</span> <span th:text="${param.password[0]}">密码</span>
            <p><font color="red">${x}将返回存储在 Thymeleaf 上下文中的变量 x 或作为请求 Request 作用范围域中的属性。</font></p><br/>
            访问 request 作用范围域中的变量：<span th:text="${book}">图书信息</span><br/>
            <p><font color="red">${session.x}将返回一个名为 x 的会话 HttpSession 作用范围域中的属性。</font></p><br/>
            访问 session 作用范围域中的变量：<span th:text="${session.school}">培训中心</span><br/>
            <p><font color="red">${application.x}将返回一个名为 x 的全局 ServletContext 上下文作用范围域中的属性。</font></p><br/>
            访问 application 作用范围域中的变量：<span th:text="${application.name}">动态页面模板</span><br/>
        </div>
    </div>
</div>
</body>
</html>
```

success1.html 中通过使用 Thymeleaf 表达式接收了请求传递的参数 loginName 和 password，并分别访问了控制器中保存在 request、session 和 application 中的三个变量。

App.java 和之前的项目一致，此处不再赘述。单击右键，从快捷菜单中选择"运行"命令运行 main 方法。

Spring Boot 项目启动后，在浏览器中输入 URL 来测试应用。

```
http://localhost:8080
```

请求会提交到 IndexController 类的 index 方法进行处理，该方法返回字符串"index"，即跳转到 templates/index.html 页面，如图 3.6 所示。

单击超链接"测试表达式访问数据"，请求进入控制器的 regexptest 方法，返回 success1.html 页面，如图 3.7 所示。

图 3.6　Thymeleaf 常用功能测试

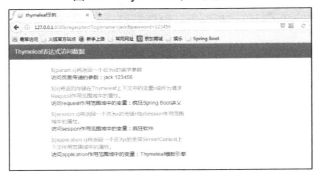

图 3.7　测试 Thymeleaf 表达式访问数据

可以看到，请求参数和保存在 request、session 和 application 作用域中的数据都访问到了。

② 测试 Thymeleaf 条件判断。

程序清单：codes/03/thymeleaftest/src/main/java/org/fkit/thymeleaftest/controller/ThymeleafController.java

```java
import javax.servlet.http.HttpServletRequest;
import javax.servlet.http.HttpSession;
import org.springframework.stereotype.Controller;
import org.springframework.web.bind.annotation.RequestMapping;
import org.springframework.web.context.request.WebRequest;

@Controller
public class ThymeleafController {

    /*
     * 将数据保存到作用域，用于测试 Thymeleaf 的条件判断
     */
    @RequestMapping("/iftest")
    public String iftest(WebRequest webRequest){
        // 将数据保存到 request 作用域，Spring MVC 更推荐使用 WebRequest
        webRequest.setAttribute("username", "fkit", webRequest.SCOPE_REQUEST);
        webRequest.setAttribute("age", 21, webRequest.SCOPE_REQUEST);
        webRequest.setAttribute("role", "admin", webRequest.SCOPE_REQUEST);
        return "success2";
    }

}
```

iftest 方法用来响应第二个请求 `<a th:href="@{iftest}">`测试条件判断`</a>`，iftest 方法中分别设置了 username、age 和 role 三个变量到 request 作用域，然后返回 success2.html。

程序清单：codes/03/thymeleaftest/src/main/resources/templates/success2.html

```html
<!DOCTYPE html>
<html xmlns:th="http://www.thymeleaf.org">
<head>
<meta charset="UTF-8"></meta>
```

```html
<title>thymeleaf 示例</title>
<link rel="stylesheet" th:href="@{css/bootstrap.min.css}" />
<link rel="stylesheet" th:href="@{css/bootstrap-theme.min.css}"/>
<script type="text/javascript" th:src="@{js/jQuery-1.11.0.min.js}"></script>
<script type="text/javascript" th:src="@{js/bootstrap.min.js}"></script>
</head>
<body>
<div class="panel panel-primary">
    <div class="panel-heading">
        <h3 class="panel-title">Thymeleaf 条件判断</h3>
    </div>
</div>
<div class="container">
    <div class="row">
        <div class="col-md-4">
            <p><font color="red">th:if 中条件成立时才显示结果</font></p><br/>
            <span th:if="${username != null}">username 不为空</span><br/>
            <span th:if="${age != null}">age 不为空</span><br/>
            <p><font color="red">th:unless 与 th:if 恰好相反，只有表达式中的条件不成立，才会显示结果</font></p><br/>
            <span th:unless="${address != null}">address 为空</span><br/>
            <p><font color="red">支持多路选择 Switch 结构，默认属性 default 可以用*表示</font></p><br/>
            <div th:switch="${role}">
              <p th:case="'admin'">管理员</p>
              <p th:case="'guest'">来宾</p>
              <p th:case="*">其他</p>
            </div>
        </div>
    </div>
</div>
</body>
</html>
```

运行项目，进入 index.html，单击超链接"测试条件判断"，请求进入控制器的 iftest 方法，返回 success2.html 页面，如图 3.8 所示。

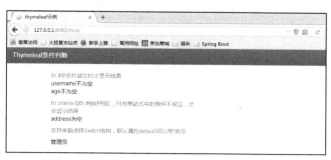

图 3.8 测试 Thymeleaf 条件判断

可以看到，条件判断 th:if、th:unless 和 th:switch 的运行结果。

③ 测试 Thymeleaf 循环取数据。

程序清单：codes/03/thymeleaftest/src/main/java/org/fkit/thymeleaftest/domain/Book.java

```java
import java.io.Serializable;
public class Book implements Serializable {
    private Integer id;
    private String title;
    private String image;
    private String author;
```

```java
    private Double price;

    public Book() {
        super();
        // TODO Auto-generated constructor stub
    }

    public Book(Integer id, String title, String image, String author, Double price) {
        super();
        this.id = id;
        this.title = title;
        this.image = image;
        this.author = author;
        this.price = price;
    }

    public Integer getId() {
        return id;
    }
    public void setId(Integer id) {
        this.id = id;
    }
    public String getTitle() {
        return title;
    }
    public void setTitle(String title) {
        this.title = title;
    }
    public String getImage() {
        return image;
    }
    public void setImage(String image) {
        this.image = image;
    }
    public String getAuthor() {
        return author;
    }
    public void setAuthor(String author) {
        this.author = author;
    }
    public Double getPrice() {
        return price;
    }
    public void setPrice(Double price) {
        this.price = price;
    }

    @Override
    public String toString() {
        return "Book [id=" + id + ", title=" + title + ", image=" + image + ", author="
                + author + ", price=" + price
                + "]";
    }
}
```

创建了一个 Book 类，并设计了 id、title、author、image 和 price 属性，用来保存图书信息，便于之后存取数据。

程序清单：codes/03/thymeleaftest/src/main/java/org/fkit/thymeleaftest/controller/ThymeleafController.java

```java
import java.util.ArrayList;
import java.util.List;
import javax.servlet.http.HttpServletRequest;
import javax.servlet.http.HttpSession;
```

```java
import org.fkit.springboot.thymeleaftest.domain.Book;
import org.springframework.stereotype.Controller;
import org.springframework.web.bind.annotation.RequestMapping;
import org.springframework.web.context.request.WebRequest;

@Controller
public class ThymeleafController {
    /*
     * 将数据保存到作用域，用于测试Thymeleaf的循环获取数据
     * */
    @RequestMapping("/eachtest")
    public String eachtest(WebRequest webRequest){
        // 模拟数据库数据保存到List集合
        List<Book> books = new ArrayList<>();
        books.add(new Book(1, "疯狂Java讲义", "java.jpg", "李刚 编著", 109.00));
        books.add(new Book(2, "轻量级Java EE企业应用实战", "ee.jpg", "李刚 编著", 108.00));
        books.add(new Book(3, "Spring+MyBatis应用实战", "SpringMyBatis.jpg", "疯狂软件 编著", 58.00));
        books.add(new Book(4, "疯狂Android讲义", "android.jpg", "李刚 编著", 108.00));
        books.add(new Book(5, "疯狂Ajax开发", "ajax.jpg", "李刚 编著", 79.00));
        // 将数据保存到request作用域
        webRequest.setAttribute("books", books, webRequest.SCOPE_REQUEST);
        return "success3";
    }

}
```

eachtest方法用来响应第三个请求<a th:href="@{eachtest}">测试循环</a>，eachtest方法中创建了5个Book对象，分别保存5本图书信息，再创建一个List集合，将所有图书信息保存到List集合，最后将List集合保存到request作用域，然后返回success3.html。

程序清单：codes/03/thymeleaftest/src/main/resources/templates/success3.html

```html
<!DOCTYPE html>
<html xmlns:th="http://www.thymeleaf.org">
<head>
<meta charset="UTF-8"></meta>
<title>thymeleaf示例</title>
<link rel="stylesheet" th:href="@{css/bootstrap.min.css}" />
<link rel="stylesheet" th:href="@{css/bootstrap-theme.min.css}"/>
<script type="text/javascript" th:src="@{js/jQuery-1.11.0.min.js}"></script>
<script type="text/javascript" th:src="@{js/bootstrap.min.js}"></script>
</head>
<body>
<div class="panel panel-primary">
    <div class="panel-heading">
        <h3 class="panel-title">Thymeleaf循环</h3>
    </div>
</div>
<div class="container">
            <div class="col-md-12">
                <div class="panel panel-primary">
                    <div class="panel-heading">
                        <h3 class="panel-title">图书信息列表</h3>
                    </div>
                    <div class="panel-body">
                        <!-- table-responsive:响应式表格，在一个表中展示所有数据，当不够显示的时候可以左右滑动浏览-->
                        <div class="table table-responsive">
                            <!--
```

```
                    .table-bordered 类为表格和其中的每个单元格增加边框。
                    .table-hover 类可以让 <tbody> 中的每一行对鼠标悬停状态做出响应。
                    -->
                    <table class="table table-bordered table-hover">
                        <thead>
                            <th class="text-center">封面</th><th class="text-center">书名</th>
                            <th class="text-center">作者</th><th class="text-center">价格</th>
                        </thead>
                        <tbody class="text-center">
                            <tr th:each="book : ${books}">
                                <td> <img src="images/java.jpg" th:src="@{'img/'+${book.image}}" height="60"/></td>
                                <td th:text="${book.title}">书名</td>
                                <td th:text="${book.author}">作者</td>
                                <td th:text="${book.price}">价格</td>
                            </tr>
                        </tbody>
                    </table>
                </div>
            </div>
        </div>
    </div>
</body>
</html>
```

运行项目，进入 index.html，单击超链接"测试循环"，请求进入控制器的 eachtest 方法，返回 success3.html 页面，如图 3.9 所示。

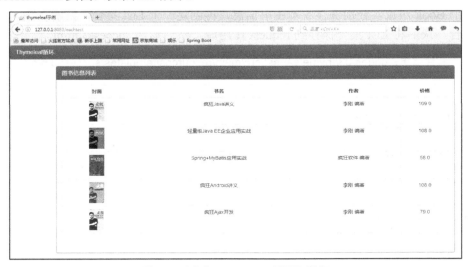

图 3.9　测试 Thymeleaf 循环取数据

可以看到，使用 th:each 可以将集合数据很方便地提取出来并显示在 html 页面上。

3.6　Spring Boot 对 JSP 的支持

Spring Boot 建议使用 HTML 来完成动态页面，但是由于大多数 Java Web 项目之前的页面展示层使用 JSP 来完成，所以 Spring Boot 同时也支持使用 JSP。

示例：Spring Boot 添加 JSP 支持

① 创建一个新的 Maven Web 项目，命名为 jsptest，如图 3.10 所示。

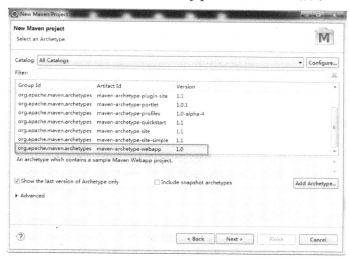

图 3.10　创建 Maven Web 项目

创建完成之后的项目结构如图 3.11 所示。

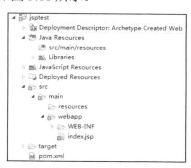

图 3.11　Maven Web 项目结构

创建完成的项目有错误，这是因为没有加入 JSP 的支持。

② 按照 Maven 项目的规范，在 src/main/下新建一个名为 resources 的文件夹，并在 src/main/resources 下再新建 static 和 templates 两个文件夹。修改 pom.xml 文件，加入 Servlet、JSTL 和 Tomcat 的依赖。再引入静态文件，操作和之前示例一致。

程序清单：codes/03/jsptest/pom.xml

```xml
<project xmlns="http://maven.apache.org/POM/4.0.0" xmlns:xsi="http://www.w3.org/2001/XMLSchema-instance"
    xsi:schemaLocation="http://maven.apache.org/POM/4.0.0 http://maven.apache.org/maven-v4_0_0.xsd">
    <modelVersion>4.0.0</modelVersion>
    <groupId>org.fkit</groupId>
    <artifactId>jsptest</artifactId>
    <packaging>war</packaging>
    <version>0.0.1-SNAPSHOT</version>
    <name>jsptest Maven Webapp</name>
    <url>http://maven.apache.org</url>

    <!-- spring-boot-starter-parent 是 Spring Boot 的核心启动器 -->
    <parent>
```

```xml
        <groupId>org.springframework.boot</groupId>
        <artifactId>spring-boot-starter-parent</artifactId>
        <version>2.0.0.RELEASE</version>
    </parent>

    <properties>
        <project.build.sourceEncoding>UTF-8</project.build.sourceEncoding>
        <project.reporting.outputEncoding>UTF-8</project.reporting.outputEncoding>
        <java.version>1.8</java.version>
    </properties>

 <dependencies>
        <!--
            添加 spring-boot-starter-web 模块依赖。
          -->
        <dependency>
            <groupId>org.springframework.boot</groupId>
            <artifactId>spring-boot-starter-web</artifactId>
        </dependency>

        <!-- 添加 Servlet 依赖 -->
        <dependency>
            <groupId>javax.servlet</groupId>
            <artifactId>javax.servlet-api</artifactId>
            <scope>provided</scope>
        </dependency>

        <!-- 添加 JSTL（JSP Standard Tag Library, JSP 标准标签库）依赖 -->
        <dependency>
            <groupId>javax.servlet</groupId>
            <artifactId>jstl</artifactId>
        </dependency>

        <!--添加 Tomcat 依赖-->
        <dependency>
            <groupId>org.springframework.boot</groupId>
            <artifactId>spring-boot-starter-tomcat</artifactId>
            <scope>provided</scope>
        </dependency>
<!-- Jasper 是 Tomcat 中使用的 JSP 引擎，运用 tomcat-embed-jasper 可以将项目与 Tomcat 分开 -->
        <dependency>
            <groupId>org.apache.tomcat.embed</groupId>
            <artifactId>tomcat-embed-jasper</artifactId>
            <scope>provided</scope>
        </dependency>

    <dependency>
      <groupId>junit</groupId>
      <artifactId>junit</artifactId>
      <scope>test</scope>
    </dependency>
  </dependencies>
  <build>
    <finalName>jsptest</finalName>
  </build>
</project>
```

③ 在 resources 下创建 application.properties 配置文件，添加页面配置信息，使之支持 JSP。

```
# 页面默认前缀目录
spring.mvc.view.prefix=/WEB-INF/jsp/
# 响应页面默认后缀
spring.mvc.view.suffix=.jsp
```

④ 创建控制器 IndexController。

程序清单：codes/03/jsptest/src/main/java/org/fkit/jsptest/controller/IndexController.java

```java
import java.util.ArrayList;
import java.util.List;
import org.fkit.jsptest.domain.Book;
import org.springframework.stereotype.Controller;
import org.springframework.ui.Model;
import org.springframework.web.bind.annotation.RequestMapping;

@Controller
public class IndexController {

    @RequestMapping("/")
    public String index(Model model){
        // 将一个 username 保存到 model
        model.addAttribute("username", "疯狂软件");
        // 模拟数据库数据保存到 List 集合
        List<Book> books = new ArrayList<>();
        books.add(new Book(1, "疯狂 Java 讲义", "java.jpg", "李刚 编著", 109.00));
        books.add(new Book(2, "轻量级 Java EE 企业应用实战", "ee.jpg", "李刚 编著", 108.00));
        books.add(new Book(3, "Spring+MyBatis 应用实战", "SpringMyBatis.jpg", "疯狂软件 编著", 58.00));
        books.add(new Book(4, "疯狂 Android 讲义", "android.jpg", "李刚 编著", 108.00));
        books.add(new Book(5, "疯狂 Ajax 开发", "ajax.jpg", "李刚 编著", 79.00));
        // 将数据保存到 model
        model.addAttribute("books", books);
        return "index";
    }
}
```

index 方法用来响应请求，在 index 方法中先创建了一个 username，保存到 model 当中。接下来创建了 5 个 Book 对象分别保存 5 本图书信息，再创建了一个 List 集合，将所有图书信息保存到 List 集合，最后将 List 集合保存到 model 当中。Book 对象和之前 thymeleaftest 项目中的一致，此处不再赘述。

⑤ 创建 JSP 页面。

application.properties 文件中指定的 JSP 文件路径是/WEB-INF/jsp/，这也是项目开发中最常用的方式。在 webapp/WEB-INF/下新建一个 jsp 目录，并重新创建一个 index.jsp 文件。

程序清单：codes/03/jsptest/src/main/webapp/WEB-INF/jsp/index.jsp

```jsp
<%@ page language="java" contentType="text/html; charset=UTF-8"
    pageEncoding="UTF-8"%>
<%@ taglib prefix="c" uri="http://java.sun.com/jsp/jstl/core" %>
<!DOCTYPE html PUBLIC "-//W3C//DTD HTML 4.01 Transitional//EN" "http://www.w3.org/TR/html4/loose.dtd">
<html>
<head>
<meta http-equiv="Content-Type" content="text/html; charset=UTF-8"></meta>
<title>Insert title here</title>
<link rel="stylesheet" href="${pageContext.request.contextPath}/css/bootstrap.min.css" />
<link rel="stylesheet" href="${pageContext.request.contextPath}/css/bootstrap-theme.min.css"/>
<script type="text/javascript" src="${pageContext.request.contextPath}/js/jQuery-1.11.0.min.js"></script>
<script type="text/javascript" src="${pageContext.request.contextPath}/js/bootstrap.min.js"></script>
</head>
<body>
<div class="panel panel-primary">
```

```html
        <!-- .panel-heading 面板头信息。 -->
        <div class="panel-heading">
            <!-- .panel-title 面板标题。 -->
            <h3 class="panel-title">Spring Boot 添加 JSP 示例</h3>
        </div>
    </div>
    <div class="container">
        <div class="row">
            <div class="col-md-4">
                <h3>欢迎[<font color="red">${requestScope.username }</font>]</h3>
            </div>
        </div>
        <div class="col-md-12">
            <div class="panel panel-primary">
                <div class="panel-heading">
                    <h3 class="panel-title">图书信息列表</h3>
                </div>
                <div class="panel-body">
                    <!-- table-responsive:响应式表格，在一个表中展示所有的数据，当不够显示的时候可以左右滑动浏览-->
                    <div class="table table-responsive">
                        <!--
                            .table-bordered 类为表格和其中的每个单元格增加边框。
                            .table-hover 类可以让 <tbody> 中的每一行对鼠标悬停状态做出响应。
                        -->
                        <table class="table table-bordered table-hover">
                            <thead>
                                <tr>
                                    <th class="text-center"> 封 面 </th ><th class="text-center">书名</th>
                                    <th class="text-center"> 作 者 </th ><th class="text-center">价格</th>
                                </tr>
                            </thead>
                            <tbody class="text-center">
                                <c:forEach items="${requestScope.books }" var="book">
                                <tr>
                                    <td> <img src="images/${book.image}" height="60"/> </td>
                                    <td>${book.title}</td>
                                    <td>${book.author}</td>
                                    <td>${book.price}</td>
                                </tr>
                                </c:forEach>
                            </tbody>
                        </table>
                    </div>
                </div>
            </div>
        </div>
    </div>
</body>
</html>
```

index.jsp 中使用了 EL 表达式和 JSTL 标签库显示页面数据。

⑥ 创建 APP 类。

> 程序清单：codes/03/jsptest/src/main/java/org/fkit/jsptest/App.java

```java
import org.springframework.boot.SpringApplication;
import org.springframework.boot.autoconfigure.SpringBootApplication;
@SpringBootApplication
public class App
{
```

```
public static void main( String[] args )
{
    // SpringApplication 用于从 main 方法启动 Spring 应用的类。
    SpringApplication.run(App.class, args);
}
```

⑦ 测试应用。

单击鼠标右键，从快捷菜单中选择"运行"命令运行 App 类的 main 方法。

Spring Boot 项目启动后，在浏览器中输入 URL 测试应用。

http://localhost:8080

请求会提交到 IndexController 类的 index 方法进行处理，该方法返回字符串"index"，即跳转到 webapp/WEB-INF/jsp/index.jsp 页面，如图 3.12 所示。

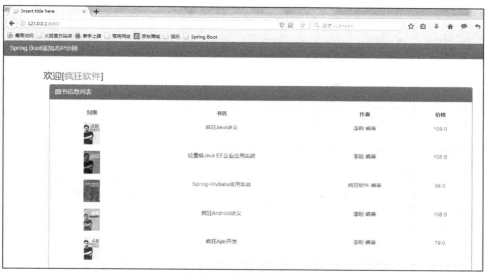

图 3.12 测试使用 JSP

虽然 Spring Boot 建议使用 HTML 作为动态页面的首选，但是也有很多开发者之前都是用 JSP 来做项目，所以 Spring Boot 也可以很完美地支持 JSP。

3.7 Spring Boot 处理 JSON 数据

在实际项目开发中，几乎所有的项目都要用到 JSON，本节将深入讲解 JSON 在 Spring Boot 中的使用。

Spring Boot 内置了 JSON 解析功能，默认使用 Jackson 来自动完成（开发中也可以把 Jackson 换成阿里的 Fastjson 或者其他 JSON 解析器），当 Controller 返回的是一个 Java 对象或者是 List 集合时，Spring Boot 自动将其转换成 JSON 数据，使用起来非常简单。

示例：Spring Boot 处理 JSON

创建一个新的 Maven 项目，命名为 jsontest。按照 Maven 项目的规范，在 src/main/下新建一个名为 resources 的文件夹，并在 src/main/resources 下新建 static 和 templates 两个文件夹。

修改 pom.xml 文件，文件内容和之前示例一致，引入静态文件，操作和之前的示例一致。

① 创建实体类。

程序清单：codes/03/jsontest/src/main/java/org/fkit/jsontest/domain/Book.java

```java
import java.io.Serializable;
public class Book implements Serializable {

    private Integer id;
    private String name;
    private String author;
    private String image;
    private Double price;
    private String remark;

    public Book() {
        super();
        // TODO Auto-generated constructor stub
    }

    public Book(Integer id, String name, String author, String image, Double price) {
        super();
        this.id = id;
        this.name = name;
        this.author = author;
        this.image = image;
        this.price = price;
    }
    // 省略 set/get 方法
}
```

Book 类是一个简单的实体类，用来保存图书信息。

② 创建 index.html。

程序清单：codes/03/jsontest/src/main/resources/templates/index.html

```html
<!DOCTYPE html>
<html xmlns:th="http://www.thymeleaf.org">
<head>
<meta charset="UTF-8"></meta>
<title>Spring Boot 自动转换 JSON 数据</title>
<link rel="stylesheet" th:href="@{css/bootstrap.min.css}" />
<link rel="stylesheet" th:href="@{css/bootstrap-theme.min.css}"/>
<script type="text/javascript" th:src="@{js/jQuery-1.11.0.min.js}"></script>
<script type="text/javascript" th:src="@{js/bootstrap.min.js}"></script>
<script type="text/javascript">
$(document).ready(function(){
    findBook();
});
function findBook(){
    $.ajax("/findBook",// 发送请求的 URL 字符串
        {
          dataType : "json", // 预期服务器返回的数据类型
            type : "post", // 请求方式 POST 或 GET
          contentType:"application/json", // 发送信息至服务器时的内容编码类型
          // 发送到服务器的数据
          data:JSON.stringify({id : 1, name : "Spring MVC 企业应用实战"}),
          async: true , // 默认设置下，所有请求均为异步请求。如果设置为 false，则发送同步请求
          // 请求成功后的回调函数
          success :function(data){
                console.log(data);
                $("#image").attr("src","images/"+data.image+"");
                $("#name").html(data.name);
                $("#author").html(data.author);
                $("#price").html(data.price);
                $("#remark").html(data.remark);
          },
```

```html
            // 请求出错时调用的函数
            error:function(){
                alert("数据发送失败");
            }
        });
    }
    </script>
</head>
<body>
<div class="panel panel-primary">
    <div class="panel-heading">
        <h3 class="panel-title">Spring Boot 中 Java 对象转换 JSON</h3>
    </div>
</div>
<div class="container">
    <div class="row">
        <div class="col-md-4">
            <p>封面：<img id="image" src="images/java.jpg"/></p>
            <p>书名：<span id="name"></span></p>
            <p>作者：<span id="author"></span></p>
            <p>价格：<span id="price"></span></p>
            <p>简介：<span id="remark"></span></p>
        </div>
    </div>
</div>
</body>
</html>
```

index.html 页面使用 jQuery 发送 JSON 数据，页面载入时调用 findBook 函数。findBook 函数发送异步请求到"/findBook"，注意加粗的代码 contentType:"application/json"，表示发送的内容编码格式为 json 类型；data:JSON.stringify({id : 1, name : "Spring MVC 企业应用实战"})，表示发送一个 JSON 数据；请求成功将返回一个 JSON 数据，接到返回的数据后将数据设置到页面的<span>当中。

③ 创建 AppController 控制器。

程序清单：codes/03/jsontest/src/main/java/org/fkit/jsontes/controller/AppController.java

```java
import org.springframework.stereotype.Controller;
import org.springframework.web.bind.annotation.RequestMapping;
@Controller
public class AppController {

    @RequestMapping(value = "/index")
    public String index() {
        return "index";
    }
}
```

④ 创建 BookController 控制器。

程序清单：codes/03/jsontest/src/main/java/org/fkit/jsontes/controller/BookController.java

```java
import java.util.ArrayList;
import java.util.List;
import org.fkit.jsontest.domain.Book;
import org.springframework.web.bind.annotation.RequestBody;
import org.springframework.web.bind.annotation.RequestMapping;
import org.springframework.web.bind.annotation.RestController;

@RestController
public class BookController {
```

```java
/**
 * Spring Boot 默认使用 Jackson 框架解析 JSON
 */
@RequestMapping("/findBook ")
public Book findBook (@RequestBody Book book){
    // 观察页面传入的 JSON 数据是否封装到 Book 对象
    System.out.println(book);
    // 设置 book 的其他信息
    book.setAuthor("肖文吉");
    book.setImage("SpringMyBatis.jpg");
    book.setPrice(58.0);
    book.setRemark("媲美于 SSH 组合的轻量级 Java EE 应用开发方式");
    return book;
}
```

findBook 方法中的第一个参数@RequestBody Book book 表示，使用@RequestBody 注解获取页面传递的 JSON 数据后，将 JSON 数据设置到对应的 Book 对象的属性当中。前台 JSP 页面的 JSON 数据中传递了 id 和 name，为了测试接收 JSON 数据，使用 System.out.println(book); 代码将接收到的 JSON 数据的 Book 对象打印在控制台上。为了测试传递 JSON 数据到 html 页面，方法中还给 Book 对象的 author、image、price 和 Remark 属性分别设置了值，最后将 Book 对象返回。此时 Spring Boot 会自动将返回的 Book 对象转换成 JSON 数据。

⑤ 测试返回的 Java 对象转换成 JSON。

App.java 和之前的项目一致，此处不再赘述。单击右键，从快捷菜单中选择"运行"命令运行 main 方法。

Spring Boot 项目启动后，在浏览器中输入 URL 来测试应用。

```
http://localhost:8080/index
```

请求会提交到 AppController 类的 index 方法进行处理，该方法返回字符串"index"，即跳转到 templates/index.html 页面。

index.html 页面载入时会发送 Ajax 请求，传递 JSON 数据，BookController 接收到请求后，@RequestBody 注解会将 JSON 数据设置到 Book 参数对应的属性当中，控制台输出如下：

```
Book [id=1, name=Spring MVC 企业应用实战, author=null, image=null, price=null, remark=null]
```

可以看到，JSON 数据传递的 id 和 name 被赋值到 Book 对象的属性当中。接下来，findBook 方法给 Book 对象的 author、image、price 和 remark 属性分别设置了值，并将 Book 对象返回。

请求响应如图 3.13 所示的界面。表示 Spring Boot 成功将 JSON 数据返回到客户端。

图 3.13　Java 对象转换成 JSON 数据

接下来测试 Spring Boot 转换集合数据。

⑥ getjson.html 页面。

> 程序清单：codes/03/jsontest/src/main/resources/templates/getjson.html

```html
<!DOCTYPE html>
<html xmlns:th="http://www.thymeleaf.org">
<head>
<meta charset="UTF-8"> </meta>
<title>Spring Boot Web 开发测试</title>
<link rel="stylesheet" th:href="@{css/bootstrap.min.css}" />
<link rel="stylesheet" th:href="@{css/bootstrap-theme.min.css}"/>
<script type="text/javascript" th:src="@{js/jQuery-1.11.0.min.js}"></script>
<script type="text/javascript" th:src="@{js/bootstrap.min.js}"></script>
<script type="text/javascript">
$(document).ready(function(){
    findBooks ();
});
function findBooks (){
    $.post("/findBooks",null,
        function(data){
        $.each(data,function(){
            var tr = $("<tr align='center'/>");
            $("<img/>").attr("src","images/"+this.image).attr("height",60).appendTo("<td/>").appendTo(tr);
            $("<td/>").html(this.name).appendTo(tr);
            $("<td/>").html(this.author).appendTo(tr);
            $("#booktable").append(tr);
        })
    },"json");
}
</script>
</head>
<body>
    <div class="panel panel-primary">
        <div class="panel-heading">
            <h3 class="panel-title">Spring Boot 中集合转换 JSON</h3>
        </div>
    </div>
    <div class="container">
        <div class="col-md-12">
            <div class="panel panel-primary">
                <!-- .panel-heading 面板头信息。 -->
                <div class="panel-heading">
                    <!-- .panel-title 面板标题。 -->
                    <h3 class="panel-title">图书信息列表</h3>
                </div>
                <div class="panel-body">
                    <div class="table table-responsive">
                        <table class="table table-bordered table-hover" id="booktable">
                            <thead>
                                <tr>
                                    <th class="text-center">封面</th>
                                    <th class="text-center">书名</th>
                                    <th class="text-center">作者</th>
                                </tr>
                            </thead>
                            <tbody class="text-center"></tbody>
                        </table>
                    </div>
                </div>
            </div>
        </div>
    </div>
```

```
        </div>
    </body>
</html>
```

getjson.html 页面使用 jQuery 发送请求，页面载入时调用 findBooks 函数。findBooks 函数发送异步请求到"/findBooks"，请求成功将返回一个 JSON 数据，包含多个书籍信息，接到返回的数据后使用 jQuery 将数据设置到页面的<table>表单当中。

⑦ 向 AppController 控制器加入代码。

程序清单：codes/03/jsontest/src/main/java/org/fkit/jsontes/controller/AppController.java

```java
@RequestMapping(value = "/getjson")
public String getjson() {
    return "getjson";
}
```

⑧ 向 BookController 控制器加入代码。

程序清单：codes/03/jsontest/src/main/java/org/fkit/jsontes/controller/BookController.java

```java
@RequestMapping("/findBooks")
public List<Book> findBooks(){
    // 创建集合
    List<Book> books = new ArrayList<Book>();
    // 添加图书对象
    books.add(new Book(1,"疯狂 Java 讲义","李刚","java.jpg",109.0));
    books.add(new Book(2,"轻量级 JavaEE 企业应用实战","李刚","ee.jpg",108.0));
    books.add(new Book(3,"Spring MVC 企业应用实战","肖文吉","SpringMyBatis.jpg",58.0));
    books.add(new Book(4,"疯狂 Android 讲义","李刚","android.jpg",108.0));
    books.add(new Book(5,"疯狂 Ajax 开发","李刚","ajax.jpg",79.0));
    // 返回集合
    return books;
}
```

Spring Boot 会将 List 集合数据转换成 JSON 格式返回到客户端。

⑨ 测试返回的集合对象转换成 JSON。

运行 App 类的 main 方法。Spring Boot 项目启动后，在浏览器中输入 URL 来测试应用。

```
http://localhost:8080/getjson
```

getjson.html 页面载入时会发送 Ajax 请求，findBooks 方法创建多个 Book 对象封装到 List 集合当中返回。

请求响应如图 3.14 所示的界面。表示 Spring Boot 成功将 JSON 数据写到客户端。

图 3.14　集合转换 JSON

3.8 Spring Boot 文件上传下载

文件上传是实际项目开发当中常用的功能之一。为了能上传文件，必须将表单的 method 设置为 POST，并将 enctype 设置为 multipart/form-data。只有在这种情况下，浏览器才会把用户所选文件的二进制数据发送给服务器。

一旦设置了 enctype 为 multipart/form-data，浏览器将采用二进制流的方式来处理表单数据，而对文件上传的处理涉及在服务器端解析原始的 HTTP 响应。2003 年，Apache 软件基金会发布了开源的 Commons FileUpload 组件，很快成为 Servlet/JSP 程序员上传文件的最佳选择。

Servlet 3.0 规范已经提供了相关方法来处理文件上传，但这种上传需要在 Servlet 中完成。而 Spring MVC 则提供了更简单的封装。

Spring MVC 为文件上传提供了直接的支持，这种支持是通过即插即用的 MultipartResolver 实现的。Spring MVC 使用 Apache Commons FileUpload 技术实现了一个 MultipartResolver 实现类：CommonsMultipartResolver。因此，Spring MVC 的文件上传还需要依赖 Apache Commons FileUpload 组件。

Spring Boot 的 spring-boot-starter-web 已经集成了 Spring MVC，所以使用 Spring Boot 完成文件上传下载更加简单。

示例：Spring Boot 文件上传

创建一个新的 Maven 项目，命名为 fileuploadtest。按照 Maven 项目的规范，在 src/main/ 下新建一个名为 resources 的文件夹，并在 src/main/resources 下新建 static 和 templates 两个文件夹。

修改 pom.xml 文件，引入 Apache Commons 依赖。

程序清单：codes/03/fileuploadtest/pom.xml

```xml
<!-- Apache Commons FileUpload 组件依赖，
由于不属于 Spring Boot, 所以需要加上版本 -->
<dependency>
    <groupId>commons-fileupload</groupId>
    <artifactId>commons-fileupload</artifactId>
    <version>1.3.3</version>
</dependency>
```

程序清单：codes/03/fileuploadtest/src/main/resources/templates/index.html

```html
<!DOCTYPE html>
<html xmlns:th="http://www.thymeleaf.org">
<head>
<meta charset="UTF-8"></meta>
<title>文件上传下载示例</title>
<link rel="stylesheet" th:href="@{css/bootstrap.min.css}" />
<link rel="stylesheet" th:href="@{css/bootstrap-theme.min.css}"/>
<script type="text/javascript" th:src="@{js/jQuery-1.11.0.min.js}"></script>
<script type="text/javascript" th:src="@{js/bootstrap.min.js}"></script>
</head>
<body>
<div class="panel panel-primary">
    <!-- .panel-heading 面板头信息。 -->
    <div class="panel-heading">
        <!-- .panel-title 面板标题。 -->
        <h3 class="panel-title">文件上传下载示例：简单文件上传</h3>
    </div>
</div>
```

```html
        <div class="container">
            <div class="row">
                <div class="col-md-8">
                    <form class="form-horizontal" action="upload" enctype="multipart/form-data" method="post">
                        <div class="form-group">
                            <div class="input-group col-md-4">
                                <span class="input-group-addon"><i class="glyphicon glyphicon-pencil"></i></span>
                                <input class="form-control" placeholder="文件描述" type="text" name="description" />
                            </div>
                        </div>
                        <div class="form-group">
                            <div class="input-group col-md-4">
                                <span class="input-group-addon"><i class="glyphicon glyphicon-search"></i></span>
                                <input class="form-control" placeholder="请选择文件" type="file" name="file"/>
                            </div>
                        </div>
                        <div class="form-group">
                            <div class="col-md-4">
                                <div class="btn-group btn-group-justified" >
                                    <div class="btn-group" >
                                        <button type="submit" class="btn btn-success" id="submitbtn">
                                            <span class="glyphicon glyphicon-share"></span> 文件上传</button>
                                    </div>
                                </div>
                            </div>
                        </div>
                    </form>
                </div>
            </div>
        </div>
</body>
</html>
```

负责上传文件的表单和一般表单有一些区别,表单的编码类型必须是"multipart/form-data"类型。

程序清单:codes/03/fileuploadtest/src/main/java/org/fkit/fileuploadtest/controller/FileUploadController.java

```java
import java.io.File;
import javax.servlet.http.HttpServletRequest;
import org.springframework.stereotype.Controller;
import org.springframework.web.bind.annotation.PostMapping;
import org.springframework.web.bind.annotation.RequestMapping;
import org.springframework.web.bind.annotation.RequestParam;
import org.springframework.web.multipart.MultipartFile;

@Controller
public class FileUploadController {

    @RequestMapping("/")
    public String index(){
        return "index";
    }

    // 上传文件会自动绑定到MultipartFile中
    @PostMapping(value="/upload")
    public String upload(HttpServletRequest request,
```

```
            @RequestParam("description") String description,
            @RequestParam("file") MultipartFile file) throws Exception{
    // 接收参数 description
    System.out.println("description = " + description);
    // 如果文件不为空，写入上传路径
    if(!file.isEmpty()){
        // 上传文件路径
        String path = request.getServletContext().getRealPath(
                "/upload/");
        System.out.println("path = " + path);
        // 上传文件名
        String filename = file.getOriginalFilename();
        File filepath = new File(path,filename);
        // 判断路径是否存在，如果不存在就创建一个
        if (!filepath.getParentFile().exists()) {
            filepath.getParentFile().mkdirs();
        }
        // 将上传文件保存到一个目标文件当中
        file.transferTo(new File(path+File.separator+ filename));
        return "success";
    }else{
        return "error";
    }
}
```

Spring MVC 会将上传文件绑定到 MultipartFile 对象中。MultipartFile 提供了获取上传文件内容、文件名等方法，通过 transferTo()方法还可以将文件存储到磁盘当中，MultipartFile 对象的重点方法如下。

➢ byte[] getBytes()：获取文件数据。
➢ String getContentType()：获取文件 MIME 类型，如 image/jpeg 等。
➢ InputStream getInputStream()：获取文件流。
➢ String getName()：获取表单中文件组件的名字。
➢ String getOriginalFilename()：获取上传文件的原名。
➢ long getSize()：获取文件的字节大小，单位为 byte。
➢ boolean isEmpty()：是否有上传的文件。
➢ void transferTo(File dest)：将上传文件保存到一个目标文件当中。

运行 App 类的 main 方法。Spring Boot 项目启动后，在浏览器中输入 URL 来测试应用。
`http://localhost:8080`

文件上传页面如图 3.15 所示。

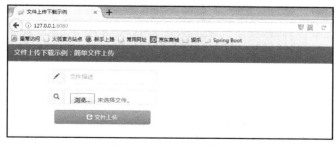

图 3.15　文件上传页面

输入文件描述信息"Spring+MyBatis 企业应用实战"，单击"浏览"按钮选择需要上传的文件图片，单击"文件上传"按钮，图片会被上传保存到项目的 upload 文件夹下，跳转到成

功页面，如图 3.16 所示。

图 3.16 文件上传成功

查看控制台输出，可以看到上传的文件描述信息和文件上传之后的保存路径即当前项目部署在 Web 服务器路径下的"upload"文件夹。进入"upload"文件夹可以看到成功上传的文件图片。

此时重新回到文件上传页面，单击"浏览"按钮选择一个比较大的图片进行上传，文件上传会失败，并且跳转到文件上传失败页面。控制台提示错误信息。

```
org.apache.tomcat.util.http.fileupload.FileUploadBase$FileSizeLimitExceededException: The field file exceeds its maximum permitted size of 1048576 bytes.
```

原因是默认上传文件大小是 1MB，当上传文件超过 1MB 时就会报错。可以在 application.properties 中设置上传文件的参数。

```
spring.http.multipart.maxFileSize=50MB
spring.http.multipart.maxRequestSize=50MB
```

spring.http.multipart.maxFileSize 用于设置单个文件大小。spring.http.multipart.maxRequestSize 用于设置总上传的数据大小。

再次运行 App 类的 main 方法。Spring Boot 项目启动后，重新上传一个较大的图片文件，此时上传成功。

示例：使用对象方式接收上传文件

在实际项目当中，很多时候上传的文件会作为对象的属性保存。Spring Boot 的处理也非常简单。

程序清单：codes/03/fileuploadtest/src/main/resources/templates/registerForm.html

```html
<!DOCTYPE html>
<html xmlns:th="http://www.thymeleaf.org">
<head>
<meta charset="UTF-8"></meta>
<title>文件上传下载示例</title>
<link rel="stylesheet" th:href="@{css/bootstrap.min.css}" />
<link rel="stylesheet" th:href="@{css/bootstrap-theme.min.css}"/>
<script type="text/javascript" th:src="@{js/jQuery-1.11.0.min.js}"></script>
<script type="text/javascript" th:src="@{js/bootstrap.min.js}"></script>
</head>
<body>
<div class="panel panel-primary">
    <!-- .panel-heading 面板头信息。 -->
    <div class="panel-heading">
        <!-- .panel-title 面板标题。 -->
        <h3 class="panel-title">文件上传下载示例：用户注册</h3>
    </div>
</div>
<div class="container">
    <div class="row">
        <div class="col-md-8">
```

```html
            <form class="form-horizontal" action="register" enctype="multipart/form-data" method="post">
                <div class="form-group">
                    <div class="input-group col-md-4">
                        <span class="input-group-addon"><i class="glyphicon glyphicon-user"></i></span>
                        <input class="form-control" placeholder="用户名" type="text" name="username" />
                    </div>
                </div>
                <div class="form-group">
                    <div class="input-group col-md-4">
                        <span class="input-group-addon"><i class="glyphicon glyphicon-search"></i></span>
                        <input class="form-control" placeholder="请选择头像" type="file" name="headPortrait"/>
                    </div>
                </div>
                <div class="form-group">
                    <div class="col-md-4">
                        <div class="btn-group btn-group-justified" >
                            <div class="btn-group" >
                                <button type="submit" class="btn btn-success" id="submitbtn">
                                <span class="glyphicon glyphicon-share"></span> 注册</button>
                            </div>
                        </div>
                    </div>
                </div>
            </form>
        </div>
    </div>
</div>
</body>
</html>
```

registerForm.html 是一个用户注册页面，需要输入用户名和上传用户头像。

程序清单：codes/03/fileuploadtest/src/main/java/org/fkit/fileuploadtest/domain/User.java

```java
import java.io.Serializable;
import org.springframework.web.multipart.MultipartFile;
public class User implements Serializable{
    private String username;
    private MultipartFile headPortrait;
    // 省略构造器和 get/set 方法……
}
```

User 是对用户信息的一个封装，属性 headPortrait 对应页面 html 的 file 类型，类型为 MultipartFile，上传文件会自动绑定到 headPortrait 属性当中。

程序清单：codes/03/fileuploadtest/src/main/java/org/fkit/fileuploadtest/controller/FileUploadController.java

```java
@RequestMapping(value="/register")
public String register(HttpServletRequest request,
    @ModelAttribute User user,
    Model model)throws Exception{
    // 接收参数 username
    System.out.println("username = " +user.getUsername());
    // 如果文件不为空，写入上传路径
    if(!user.getHeadPortrait().isEmpty()){
        // 上传文件路径
        String path = request.getServletContext().getRealPath(
            "/upload/");
```

```
            System.out.println("path = " + path);
            // 上传文件名
            String filename = user.getHeadPortrait().getOriginalFilename();
            File filepath = new File(path,filename);
            // 判断路径是否存在，如果不存在就创建一个
            if (!filepath.getParentFile().exists()) {
                filepath.getParentFile().mkdirs();
            }
            // 将上传文件保存到一个目标文件当中
            user.getHeadPortrait().transferTo(new File(path+File.separator+ filename));
            // 将用户添加到model
            model.addAttribute("user", user);
            return "userInfo";
        }else{
            return "error";
        }
    }
```

register 方法使用@ModelAttribute 注解将表单参数绑定到 User 对象，html 控件的 username 会保存到 User 对象的 username 属性，html 控件的 headPortrait 会保存到 User 对象的 headPortrait 属性，转换为 MultipartFile 类型。文件上传成功后将用户信息保存到 model 当中，跳转到 userInfo.html。

程序清单：codes/03/fileuploadtest/src/main/resources/templates/userInfo.html

```
<!DOCTYPE html>
<html xmlns:th="http://www.thymeleaf.org">
<head>
<meta charset="UTF-8"></meta>
<title>文件上传下载示例</title>
<link rel="stylesheet" th:href="@{css/bootstrap.min.css}" />
<link rel="stylesheet" th:href="@{css/bootstrap-theme.min.css}"/>
<script type="text/javascript" th:src="@{js/jQuery-1.11.0.min.js}"></script>
<script type="text/javascript" th:src="@{js/bootstrap.min.js}"></script>
</head>
<body>
<div class="panel panel-primary">
    <!-- .panel-heading 面板头信息。 -->
    <div class="panel-heading">
        <!-- .panel-title 面板标题。 -->
        <h3 class="panel-title">文件上传下载示例：文件下载</h3>
    </div>
</div>
<div class="container">
    <div class="panel panel-primary">
        <!-- .panel-heading 面板头信息。 -->
        <div class="panel-heading">
            <!-- .panel-title 面板标题。 -->
            <h3 class="panel-title">用户信息列表</h3>
        </div>
        <div class="panel-body">
            <div class="table table-responsive">
                <table class="table table-bordered table-hover" >
                    <tbody class="text-center">
                        <tr>
                            <td><img th:src="@{'upload/'+${user.headPortrait.originalFilename}}" height="30"/>s</td>
                            <td th:text="${user.username}">用户名</td>
                            <td >
                                <a th:href="@{download(filename=${user.headPortrait.originalFilename })}">下载头像</a>
                            </td>
```

```
                </tr>
            </tbody>
        </table>
    </div>
   </div>
  </div>
 </body>
</html>
```

userInfo.html 显示用户信息。

再次运行 App 类的 main 方法。Spring Boot 项目启动后，在浏览器中输入 URL 来测试应用。

```
http://localhost:8080/registerForm
```

文件上传页面如图 3.17 所示。

图 3.17　用户注册页面

输入用户名，选择上传的用户图像，单击"注册"按钮，注册用户信息并保存用户图像，跳转到用户信息页面，如图 3.18 所示。

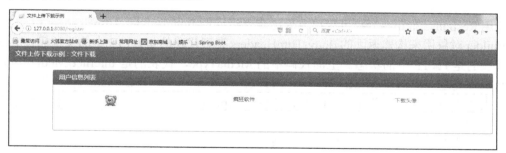

图 3.18　下载用户头像页面

示例：文件下载

单击图 3.18 中的"下载头像"超链接，可以进入 download 控制器进行文件下载操作。文件下载比较简单，直接在页面上给出一个超链接，该链接的 href 属性等于要下载文件的文件名，就可以实现文件下载了。但是如果该文件的文件名为中文，在某些早期的浏览器上会导致下载失败，如果使用最新的 Firefox、Opera、Chrome、Safari，都可以正常下载文件名为中文的文件。

Spring MVC 提供了一个 ResponseEntity 类型，可以很方便地定义返回的 HttpHeaders 和 HttpStatus。

程序清单：codes/03/fileuploadtest/src/main/java/org/fkit/fileuploadtest/controller/FileUploadController.java

```
@RequestMapping(value="/download")
public ResponseEntity<byte[]> download(HttpServletRequest request,
    @RequestParam("filename") String filename,
```

```
        @RequestHeader("User-Agent") String userAgent,
        Model model)throws Exception{
    // 下载文件路径
    String path = request.getServletContext().getRealPath(
    "/upload/");
    // 构建 File
    File file = new File(path+File.separator+ filename);
        // ok 表示 HTTP 中的状态 200
        BodyBuilder builder = ResponseEntity.ok();
        // 内容长度
        builder.contentLength(file.length());
        // application/octet-stream： 二进制流数据（最常见的文件下载）
        builder.contentType(MediaType.APPLICATION_OCTET_STREAM);
        // 使用 URLDecoder.decode 对文件名进行解码
        filename = URLEncoder.encode(filename, "UTF-8");
        // 设置实际的响应文件名，告诉浏览器文件要用于"下载"和"保存"
        // 不同的浏览器，处理方式不同，要根据浏览器版本区别对待
        if (userAgent.indexOf("MSIE") > 0) {
            // 如果是 IE，只需要用 UTF-8 字符集进行 URL 编码即可
            builder.header("Content-Disposition", "attachment; filename=" + filename);
        } else {
            // 而 FireFox、Chrome 等浏览器，则需要说明编码的字符集
            // 注意 filename 后面有个*号，在 UTF-8 后面有两个单引号
            builder.header("Content-Disposition", "attachment; filename*=UTF-8''" + filename);
        }
        return builder.body(FileUtils.readFileToByteArray(file));
    }
```

download 处理方法接收到页面传递的文件名 filename 后，使用 Apache Commons FileUpload 组件的 FileUtils 读取项目的 upload 文件夹下的该文件，并将其构建成 ResponseEntity 对象返回客户端下载。

使用 ResponseEntity 对象，可以很方便地定义返回的 BodyBuilder、HttpHeaders 和 HttpStatus。BodyBuilder 对象用来构建返回的 Body；HttpHeaders 类型代表的是 HTTP 中的头信息；HttpStatus 类型代表的是 HTTP 中的状态。上面代码中的 MediaType，代表的是 Internet Media Type，即互联网媒体类型，也叫作 MIME 类型。在 HTTP 消息头中，使用 Content-Type 来表示具体请求中的媒体类型信息。有关 BodyBuilder、MediaType 和 HttpStatus 类的详细信息参考 Spring MVC 的 API 文档。

单击下载页面的超链接，显示文件下载窗口，如图 3.19 所示。

图 3.19　文件下载窗口

单击"浏览"按钮，选择下载文件保存路径，单击"确定"按钮，文件下载完成。

3.9 Spring Boot 的异常处理

在互联网时代，我们所开发的应用大多是直面用户的，程序中的任何一点小疏忽都可能导致用户的流失，而程序出现异常往往又是不可避免的，那该如何减少程序异常对用户体验的影响呢？其实方法很简单，对异常进行捕获，然后给予相应的处理。下面介绍 Spring Boot 提供的异常处理方式。

示例：ExceptionHandler 处理异常

创建一个新的 Maven 项目，命名为 exceptiontest。按照 Maven 项目的规范，在 src/main/下新建一个名为 resources 的文件夹，并在 src/main/resources 下新建 static 和 templates 两个文件夹。

修改 pom.xml 文件，文件内容和之前示例一致，引入静态文件，操作和之前示例一致。

程序清单：codes/03/exceptiontest/src/main/resources/exceptiontest/index.html

```html
<!DOCTYPE html>
<html xmlns:th="http://www.thymeleaf.org">
<head>
<meta charset="UTF-8"></meta>
<title>异常处理示例</title>
<link rel="stylesheet" th:href="@{css/bootstrap.min.css}" />
<link rel="stylesheet" th:href="@{css/bootstrap-theme.min.css}"/>
<script type="text/javascript" th:src="@{js/jQuery-1.11.0.min.js}"></script>
<script type="text/javascript" th:src="@{js/bootstrap.min.js}"></script>
</head>
<body>
<div class="panel panel-primary">
    <!-- .panel-heading 面板头信息。 -->
    <div class="panel-heading">
        <!-- .panel-title 面板标题。 -->
        <h3 class="panel-title">异常处理示例</h3>
    </div>
</div>
<div class="container">
    <div class="row">
        <div class="col-md-4">
            <a th:href="@{hello}">Spring Boot 默认异常处理</a><br/><br/>
            <a th:href="@{test}">@ExceptionHandler 处理异常</a><br/><br/>
        </div>
    </div>
</div>
</body>
</html>
```

index.html 中提供了两个超链接，hello 超链接请求抛出异常不做任何处理，test 超链接请求使用@ExceptionHandler 注解处理异常。

程序清单：codes/03/exceptiontest/src/main/java/org/fkit/exceptiontest/controller/TestController.java

```java
import org.springframework.stereotype.Controller;
import org.springframework.web.bind.annotation.ExceptionHandler;
import org.springframework.web.bind.annotation.RequestMapping;

@Controller
public class TestController{

    @RequestMapping("/")
    public String index(){
```

```
        return "index";
    }

    @RequestMapping("/hello")
    public String hello() throws Exception{
        // 抛出异常
        throw new Exception();
    }
}
```

运行 App 类的 main 方法。Spring Boot 项目启动后，在浏览器中输入 URL 来测试应用。

`http://localhost:8080`

请求会提交到 TestController 类的 index 方法进行处理，该方法返回字符串"index"，即跳转到 index.html 页面，如图 3.20 所示。

图 3.20　异常处理示例

单击"Spring Boot 默认异常处理"，发送"hello"请求，TestController 的 hello 方法抛出一个异常，此时异常被 Spring Boot 捕获，默认跳转到 error 页面，如果没有 error 页面，则显示默认异常信息，如图 3.21 所示。

图 3.21　默认异常信息

接下来，使用@ExceptionHandler 注解来处理异常。

程序清单：codes/03/exceptiontest/src/main/java/org/fkit/exceptiontest/controller/TestController.java

```
import org.springframework.stereotype.Controller;
import org.springframework.web.bind.annotation.ExceptionHandler;
import org.springframework.web.bind.annotation.RequestMapping;
@RequestMapping("/test")
public String test() throws Exception{
    System.out.println("test()......");
    // 模拟异常
    int i = 5/0;
    return "success";
}
/**
 * 在异常抛出的时候,Controller 会使用@ExceptionHandler 注解处理异常,而不会抛给 Servlet 容器
 * */
@ExceptionHandler(value = Exception.class)
public String testErrorHandler(Exception e)  {
    System.out.println("TestController testErrorHandler()......");
```

```
            return "服务器故障，请联系管理员。";
    }
```

@ExceptionHandler 用来注解处理异常的方法，value 属性表示处理的异常类型。如果在一个 Controller 中有一个用@ExceptionHandler 修饰的方法，当 Controller 的任何一个方法抛出异常时，都由@ExceptionHandler 注解修饰的方法处理异常，而不会抛给 Servlet 容器。

再次运行 App 类的 main 方法。Spring Boot 项目启动后，进入 index.html 页面，单击 "@ExceptionHandler 处理异常" 超链接，发送 "test" 请求，进入 TestController 的 test 方法，该方法模拟了一个异常，当异常抛出时，会使用@ExceptionHandler 注解的 testErrorHandler 方法进行处理，如图 3.22 所示。

图 3.22 @ExceptionHandler 处理异常

同时，观察控制台输出。

```
TestController testErrorHandler()......
```

看到输出结果，说明抛出异常时进入了@ExceptionHandler 注解修饰的 testErrorHandler 方法。

再次单击 "Spring Boot 默认异常处理" 超链接，观察控制台输出，说明 "hello" 请求抛出异常时也被@ExceptionHandler 注解修饰的 testErrorHandler 方法捕获了。

> **注意**
> 现在显示异常信息的页面还是 Spring Boot 默认的处理页面，因为在 testErrorHandler 方法中只是简单地打印了一句话而已，并没有提供显示错误信息的自定义页面。

示例：父类 Controller 处理异常

项目往往存在着多个 Controller，而它们在异常处理方面又存在着很多的共性，这样就不太适合在每一个 Controller 里面都编写一个对应的异常处理方法。可以将异常处理方法向上移到父类中，然后所有的 Controller 统一继承父类即可。

程序清单：codes/03/exceptiontest/src/main/java/org/fkit/exceptiontest/controller/BaseController.java

```java
import javax.servlet.http.HttpServletRequest;
import org.springframework.web.bind.annotation.ExceptionHandler;
import org.springframework.web.servlet.ModelAndView;
public class BaseController {
    @ExceptionHandler(value = Exception.class)
    public ModelAndView defaultErrorHandler(HttpServletRequest req, Exception e) throws Exception {
        System.out.println("BaseController defaultErrorHandler()......");
        ModelAndView mav = new ModelAndView();
        mav.addObject("exception", e);
        mav.addObject("url", req.getRequestURL());
        mav.setViewName("error");
```

```
        return mav;
    }
}
```

BaseController 类当中定义了一个 @ExceptionHandler 注解修饰的处理异常的 defaultErrorHandler 方法，该方法中使用 ModelAndView 保存了 url 和 exception 两个属性，然后跳转到 error 页面。

程序清单：codes/03/exceptiontest/src/main/java/org/fkit/exceptiontest/controller/UserController.java

```java
import org.springframework.stereotype.Controller;
import org.springframework.web.bind.annotation.RequestMapping;
import org.springframework.stereotype.Controller;
import org.springframework.web.bind.annotation.RequestMapping;

@Controller
public class UserController extends BaseController{

    @RequestMapping("/login")
    public String login(String username) throws Exception{
        System.out.println("login()……");
        if(username == null ){
            throw new NullPointerException("用户名不存在!");
        }
        return "success";
    }
}
```

UserController 继承 BaseController，当抛出异常时，会使用 BaseController 中定义的处理异常的方法。

程序清单：codes/03/exceptiontest/src/main/java/org/fkit/exceptiontest/controller/BookController.java

```java
import org.springframework.stereotype.Controller;
import org.springframework.web.bind.annotation.RequestMapping;
import org.springframework.stereotype.Controller;
import org.springframework.web.bind.annotation.RequestMapping;

@Controller
public class BookController extends BaseController{

    @RequestMapping("/find")
    public String find() throws Exception{
        System.out.println("find()……");
        int i = 5/0;
        return "success";
    }
}
```

BookController 继承 BaseController，当抛出异常时，会使用 BaseController 中定义的处理异常的方法。

程序清单：codes/03/exceptiontest/src/main/resources/exceptiontest/index.html

```html
<div class="container">
    <div class="row">
        <div class="col-md-4">
            <a th:href="@{hello}">Spring Boot 默认异常处理</a><br/><br/>
            <a th:href="@{test}">@ExceptionHandler 处理异常</a><br/><br/>
            <hr/>
            <a th:href="@{login}">UserController：父级 Controller 异常处理</a><br/><br/>
            <a th:href="@{find}">BookController：父级 Controller 异常处理</a><br/><br/>
```

在 index.html 中增加两个超链接，用来测试父类 Controller 处理异常。

程序清单：codes/03/exceptiontest/src/main/resources/exceptiontest/error.html

```html
<!DOCTYPE html>
<html xmlns:th="http://www.thymeleaf.org">
<head>
<meta charset="UTF-8"></meta>
<title>异常处理示例</title>
<link rel="stylesheet" th:href="@{css/bootstrap.min.css}" />
<link rel="stylesheet" th:href="@{css/bootstrap-theme.min.css}"/>
<script type="text/javascript" th:src="@{js/jQuery-1.11.0.min.js}"></script>
<script type="text/javascript" th:src="@{js/bootstrap.min.js}"></script>
</head>
<body>
<div class="panel panel-primary">
    <!-- .panel-heading 面板头信息。 -->
    <div class="panel-heading">
        <!-- .panel-title 面板标题。 -->
        <h3 class="panel-title">异常处理示例</h3>
    </div>
</div>
<div class="container">
    <div class="row">
        <div class="col-md-4">
            <div th:text="统一异常处理"></div>
            <div th:text="${url}"></div>
            <div th:text="${exception.message}"></div>
        </div>
    </div>
</div>
</body>
</html>
```

error.html 作为本示例的自定义异常处理页面，只是简单打印了异常的请求路径和异常信息，在实际项目开发中异常页面可以个性化定制。

再次运行 App 类的 main 方法。Spring Boot 项目启动后，进入 index.html 页面，单击"UserController：父级 Controller 异常处理"超链接，发送"login"请求，进入 UserController 的 login 方法，该方法接收一个参数，当参数为空时抛出异常，会被父类使用@ExceptionHandler 注解的 defaultErrorHandler 方法进行处理，如图 3.23 所示。

图 3.23 父类 Controller 处理异常

单击"BookController：父级 Controller 异常处理"超链接，发送"find"请求，进入 BookController 的 login 方法，该方法中模拟了一个异常，当抛出异常时，会被父类使用@ExceptionHandler 注解的 defaultErrorHandler 方法进行处理，如图 3.24 所示。

图 3.24　父类 Controller 处理异常

示例：Advice 处理异常返回 JSON

使用父级 Controller 完成异常处理也有其自身的缺点，那就是代码耦合严重，一旦哪天忘记继承 BaseController，异常就没有进行处理而直达客户了。想要解除这种耦合关系，可以使用 @ControllerAdvice 进行处理。

@ControllerAdvice 注解的类是当前项目中所有类的统一异常处理类，@ExceptionHandler 注解的方法用来定义函数针对的异常类型以及异常如何处理，不用在每个 Controller 中逐一定义异常处理方法。

前面的示例当中使用 ModelAndView 保存属性返回的异常信息，在很多时候，需要返回 JSON 数据，这在 Spring Boot 中操作起来也非常简单，只需在@ExceptionHandler 之后加入@ResponseBody，就能将处理方法返回的 Map 集合转换为 JSON 格式。

程序清单：codes/03/exceptiontest/src/main/java/org/fkit/exceptiontest/exception/GlobalExceptionHandler.java

```java
import javax.servlet.http.HttpServletRequest;
import org.springframework.web.bind.annotation.ControllerAdvice;
import org.springframework.web.bind.annotation.ExceptionHandler;
import org.springframework.web.bind.annotation.ResponseBody;
import java.util.HashMap;
import java.util.Map;

@ControllerAdvice
public class GlobalExceptionHandler {

    @ExceptionHandler(value = Exception.class)
    @ResponseBody
    public Object globalErrorHandler(HttpServletRequest request,Exception e) throws Exception {
        System.out.println("GlobalExceptionHandler globalErrorHandler()……");
        // 创建返回对象 Map 并设置属性，会被@ResponseBody 注解转换为 JSON 返回
        Map<String, Object> map = new HashMap<>();
        map.put("code", 100);
        map.put("message", e.getMessage());
        map.put("url", request.getRequestURL().toString());
        map.put("data", "请求失败");
        return map;
    }

}
```

最后，写一个测试 Controller。

程序清单：codes/03/exceptiontest/src/main/java/org/fkit/exceptiontest/controller/DeptController.java

```java
import org.springframework.stereotype.Controller;
import org.springframework.web.bind.annotation.RequestMapping;

@Controller
public class DeptController {

    @RequestMapping("/add")
    public String add(String deptname) throws Exception{
        System.out.println("add()......");
        if(deptname == null ){
            throw new NullPointerException("部门名称不能为空！");
        }
        return "success";
    }
}
```

程序清单：codes/03/exceptiontest/src/main/resources/exceptiontest/index.html

```html
<div class="container">
    <div class="row">
        <div class="col-md-4">
            <a th:href="@{hello}">Spring Boot 默认异常处理</a><br/><br/>
            <a th:href="@{test}">@ExceptionHandler 处理异常</a><br/><br/>
            <hr/>
            <a th:href="@{login}">UserController：父级 Controller 异常处理</a><br/><br/>
            <a th:href="@{find}">BookController：父级 Controller 异常处理</a><br/><br/>
            <hr/>
            <a th:href="@{add}">Advice 异常处理返回 json</a><br/><br/>
        </div>
    </div>
</div>
```

在 index.html 中增加一个超链接。

再次运行 App 类的 main 方法。Spring Boot 项目启动后，进入 index.html 页面，单击"Advice 异常处理返回 json"超链接，发送"add"请求，进入 DeptController 的 add 方法，该方法接收一个参数，当参数为空时抛出异常，会被@ControllerAdvice 注解修饰的全局异常处理类使用 @ExceptionHandler 注解的 globalErrorHandler 方法处理，异常信息被封装到返回对象 Result，并转为 JSON 返回，如图 3.25 所示。

图 3.25 Advice 处理异常返回 JSON

以上几种异常处理方式中，毫无疑问，@ControllerAdvice 是最方便的，所以在实际开发当中，@ControllerAdvice 也是用得最多的。而且异常处理的返回结果非常灵活，如果返回的是 View（视图），方法的返回值是 ModelAndView；如果返回的是 String 或者 JSON 数据，那么只需要在方法上添加@ResponseBody 注解就可以了。

3.10 本章小结

本章主要介绍了 Spring Boot 的 Web 开发，和 Spring Boot 建议使用的 Thymeleaf 模板引擎，包括 Thymeleaf 的基础语法、常用功能。使用 Thymeleaf 模板引擎，可以很方便地和 Spring MVC 集成，并便捷地操作 html 页面的数据。

本章还介绍了 Spring Boot 对 JSP 的支持、对 JSON 数据的处理、文件上传下载和异常处理等 Web 开发常用的功能。Spring Boot 的 Web 开发其实还使用了 Spring MVC 框架，关于 Spring MVC 的更多知识，请参考疯狂软件体系中的《Spring+MyBatis 企业应用实战》。

第 4 章将重点介绍 Spring Boot 的数据访问。

CHAPTER 4

第 4 章
Spring Boot 的数据访问

本章要点

- JPA/Hibernate/Spring Data JPA 概念
- Spring Data JPA 访问数据库
- Spring Data JdbcTemplate 访问数据库
- Spring Boot 集成 MyBatis

本章将详细介绍 Spring Boot 访问数据库的技术。Java 访问数据库经历了几个阶段，第一个阶段是直接通过 JDBC 访问，这种方式工作量极大，并且会做大量的重复劳动。之后出现了一些成熟的 ORM 框架，如 MyBatis、Hibernate 等，这些框架封装了大量数据库的访问操作，但是开发者依然要对这些框架进行二次封装。现在 Spring Data 提供了一套 JPA 接口可以非常简单地实现基于关系型数据库的访问操作。Spring Boot 提供的 Spring Data JPA 等于在 ORM 框架之上又进行了一次封装，但对数据库的具体访问依然要依赖于底层的 ORM 框架，Spring Data JPA 默认是通过 Hibernate 实现的。

4.1 Hibernate/JPA/Spring Data JPA 的概念

▶▶ 4.1.1 对象/关系数据库映射（ORM）

ORM 的全称是 Object/Relation Mapping，即对象/关系型数据库映射。可以将 ORM 理解成一种规范，它概述了这类框架的基本特征，完成面向对象的编程语言到关系型数据库的映射。当 ORM 框架完成映射后，程序员既可以利用面向对象程序设计语言的简单易用性，又可以利用关系型数据库的技术优势。因此可以把 ORM 当成应用程序和数据库的桥梁。

当使用一种面向对象的编程语言来进行应用开发时，从项目一开始就采用的是面向对象分析、面向对象设计、面向对象编程，但到了持久层数据库访问时，又必须重返关系型数据库的访问方式，这是一种非常糟糕的感觉。于是人们需要一种工具，它可以把关系型数据库包装成面向对象的模型，这个工具就是 ORM。

ORM 框架是面向对象程序设计语言与关系型数据库发展不同步时的中间解决方案。随着面向对象数据库的发展，其理论逐步完善，最终面向对象数据库会取代关系型数据库。只是这个过程不可一蹴而就，ORM 框架在此期间会蓬勃发展。但随着面向对象数据库的广泛使用，ORM 工具会逐渐消亡。

对于时下所有流行的编程语言而言，面向对象的程序设计语言代表了目前程序设计语言的主流和趋势，其具备非常多的优势。比如：

> 面向对象地建模、操作。
> 多态、继承。
> 摒弃难以理解的过程。
> 简单易用，易理解。

但数据库的发展并未能与程序设计语言同步，而且关系型数据库系统的某些优势也是面向对象语言目前无法比拟的。比如：

> 大量数据查找、排序。
> 集合数据连接操作、映射。
> 数据库访问的并发、事务。
> 数据库的约束、隔离。

面对这种面向对象语言与关系型数据库系统并存的局面，采用 ORM 就变成一种必然。只要依然采用面向对象程序设计语言，底层依然采用关系型数据库，中间就少不了 ORM 工具。采用 ORM 框架之后，应用程序不再直接访问底层数据库，而是以面向对象的方式来操作持久化对象（例如创建、修改、删除等），而 ORM 框架则将这些面向对象的操作转换成底层的 SQL 操作。

图 4.1 显示了 ORM 工具工作的示意图。

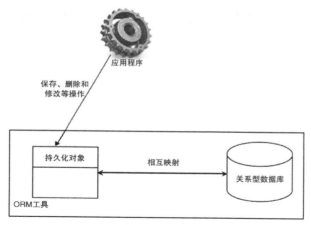

图 4.1　ORM 工具工作的示意图

正如图 4.1 所示，ORM 工具的唯一作用就是：把对持久化对象的保存、修改、删除等操作，转换成对数据库的操作。由此，程序员可以以面向对象的方式操作持久化对象，而 ORM 框架则负责将相关操作转换成对应的 SQL（结构化查询语言）操作。

▶▶ 4.1.2　基本映射方式

ORM 工具提供了持久化类和数据表之间的映射关系，通过这种映射关系的过渡，程序员可以很方便地通过持久化类实现对数据表的操作。实际上，所有的 ORM 工具大致都遵循相同的映射思路。ORM 有如下几条基本映射关系。

- **数据表映射类**。持久化类被映射到一个数据表。程序使用这个持久化类来创建实例、修改属性、删除实例时，系统会自动转换为对这个表进行 CRUD 操作。图 4.2 显示了这种映射关系。

图 4.2　数据表对应 Model 类

正如图 4.2 所示，受 ORM 管理的持久化类（就是一个普通 Java 类）对应一个数据表，只要程序对这个持久化类进行操作，系统就可以将其转换成对对应数据库表的操作。

- **数据表的行映射对象（即实例）**。持久化类会生成很多实例，每个实例对应数据表中的一行记录。当程序在应用中修改持久化类的某个实例时，ORM 工具会将其转换成对对应数据表中特定行的操作。每个持久化对象对应数据表的一行记录的示意图如图 4.3 所示。
- **数据表的列（字段）映射对象的属性**。当程序修改某个持久化对象的指定属性时（持久化实例映射到数据行），ORM 会将其转换成对对应数据表中指定数据行、指定列的操作。数据表的列被映射到对象属性的示意图如图 4.4 所示。

基于这种基本的映射方式，ORM 工具可完成对象模型和关系模型之间的相互映射。由此可见，在 ORM 框架中，持久化对象是一种媒介，应用程序只需操作持久化对象，ORM 框架则负责将这种操作转换为底层数据库操作。这种转换对开发者透明，开发者无须关心内部细节，从而将开发者从关系模型中解放出来，使得开发者能以面向对象的思维操作关系型数据库。

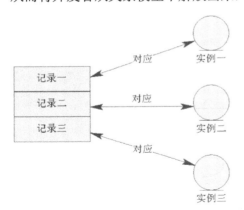

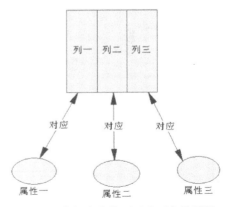

图 4.3　数据表中的行记录对应持久化对象　　　　图 4.4　数据表中的列对应对象的属性

▶▶ 4.1.3　流行的 ORM 框架简介

目前 ORM 框架的产品非常多，除了各大公司、组织的产品外，甚至一些小团队也推出自己的 ORM 框架。目前流行的 ORM 框架有如下这些产品。

➤ Hibernate 是一个开放源代码的对象关系映射框架，它对 JDBC 进行了非常轻量级的对象封装，它将 POJO 与数据库表建立映射关系，是一个全自动的 ORM 框架，Hibernate 可以自动生成 SQL 语句，自动执行，使得 Java 程序员可以随心所欲地使用面向对象编程思维来操纵数据库。Hibernate 可以应用在任何使用 JDBC 的场合，既可以在 Java 的客户端程序中使用，也可以在 Servlet/JSP 的 Web 应用中使用。最具革命意义的是，Hibernate 可以在应用 EJB 的 Java EE 架构中取代 CMP，完成数据持久化的重任。

➤ JPA 全称 Java Persistence API，是官方提出的 Java 持久化规范。JPA 通过 JDK 5.0 注解或 XML 描述对象—关系表的映射关系，并将运行期的实体对象持久化到数据库中。它为 Java 开发人员提供了一种对象—关系映射工具来管理 Java 应用中的关系数据。持久化（Persistence），即把数据（如内存中的对象）保存到可永久保存的存储设备中（如磁盘）。持久化的主要应用是将内存中的对象存储在数据库中，或者存储在磁盘文件、XML 数据文件中等，JDBC 就是一种持久化机制，文件 IO 也是一种持久化机制。

所谓的"规范"意指明文规定或约定俗成的标准，"持久化规范"就是 Java 针对持久化这一层操作指定的规范，这意味着类似于 Hibernate 这样的框架都应该遵循此规范设计，假如没有指定 JPA 规范，那么新的 ORM 框架就随意按照自己的标准来设计了，那开发人员就没法把精力全部集中在业务层上，而是在想如何学习使用各种各样的持久层框架。有了 JPA 以后，开发人员可以通过面向 JPA 开发来操作数据库,底层的 ORM 框架实现就可以实现自由切换了。JPA 的主要实现有 Hibernate、EclipseLink 和 OpenJPA 等。

Spring Data 是 Spring 的一个子项目，用于简化数据库访问，包括 NoSQL 非关系型数据库，另外还包括对关系型数据库的访问支持。Spring Data 使我们可以快速简单地使用普通的数据访问技术及新的数据访问技术，Spring Data 会让数据的访问变得更加方便。

Spring Data JPA 可以极大地简化 JPA 的写法，可以在几乎不用写实现的情况下，实现对数据的访问和操作。除了 CRUD 外，还包括如分页、排序等一些常用的功能。Spring Data 是一个开源框架，Spring Data JPA 只是 Spring Data 框架中的一个模块，所以称为 Spring Data JPA。如果单独使用 JPA 开发，你会发现代码量和使用 JDBC 开发一样有点惊人，Spring Data JPA 的出现就是为了简化 JPA 的写法，让开发者只需要编写一个接口、继承一个类就能实现 CRUD 操作。

4.2 Spring Data JPA

4.2.1 Spring Data 核心数据访问接口

Spring Data JPA 可以极大地简化 JPA 的写法，在几乎不用写接口实现的情况下完成对数据的访问和操作，由于 Spring Data JPA 是 Spring Data 下的一个模块，所以需要先掌握 Spring Data 项目的相关技术。Spring Data 项目提供了访问操作数据的统一规范，该规范约定了对于关系型和非关系型数据库操作的统一标准，主要包含 CRUD（增加、查询、修改、删除）操作、相关参数查询、分页操作、排序操作等。

Spring Data 通过提供 Repository 接口来约定数据访问的统一标准，Repository 接口的源码如下。

程序清单：org/springframework/data/repository.java
```
public interface Repository<T, ID extends Serializable> {
}
```

从源码中可以看出，Repository<T，ID extends Serializable>接口接收当前所操作的实体类型参数，以及 ID 类型参数。Repository 接口下包含了一些常用的子接口：

- CrudRepository<T，ID extends Serializable>
- PagingAndSortingRepository<T，ID extends Serializable>
- JpaRepository<T，ID extends Serializable>

在进行 Spring Boot 项目的开发中，开发者只需定义自己项目的数据访问接口，然后实现 Spring Data 提供的这些接口，就可以实现对数据的 CRUD 操作了，这也正是 Spring Data 可以简化数据访问的关键所在。接下来详细介绍这些接口。

CrudRepository

CrudRepository 提供了最基本的对实体类的增删改查操作。CrudRepository 接口的源码如下。

程序清单：org/springframework/data/repository/CrudRepository.java
```java
package org.springframework.data.repository;
import java.io.Serializable;

@NoRepositoryBean
public interface CrudRepository<T, ID extends Serializable> extends Repository<T, ID> {
    <S extends T> S save(S entity);
    <S extends T> Iterable<S> save(Iterable<S> entities);
    T findOne(ID id);
    boolean exists(ID id);
    Iterable<T> findAll();
    Iterable<T> findAll(Iterable<ID> ids);
    long count();
    void delete(ID id);
    void delete(T entity);
    void delete(Iterable<? extends T> entities);
    void deleteAll();
}
```

该接口提供了很多常用的方法。
- \<S extends T\> S save(S entity)：保存单个实体对象数据。
- \<S extends T\> Iterable\<S\> save(Iterable\<S\> entities)：保存给定的所有实体对象。
- T findOne(ID id)：根据 id 查询实体。
- boolean exists(ID id)：根据 id 判断实体是否存在。
- Iterable\<T\> findAll()：查询所有实体。
- Iterable\<T\> findAll(Iterable\<ID\> ids)：根据所提供的这些 id，将对应的实体全部查询出来。
- long count()：统计实体总数量。
- void delete(ID id)：根据 id 删除实体。
- void delete(T entity)：删除一个实体。
- void delete(Iterable\<? extends T\> entities)：删除一个实体的集合。
- void deleteAll()：删除所有实体。

示例：CrudRepository 接口访问数据

创建一个新的 Maven 项目，命名为 crudrepositorytest。按照 Maven 项目的规范，在 src/main/ 下新建一个名为 resources 的文件夹。

① 修改 pom.xml 文件。

在 pom.xml 文件中增加 Web 开发依赖配置。

```xml
<dependency>
    <groupId>org.springframework.boot</groupId>
    <artifactId>spring-boot-starter-web</artifactId>
</dependency>
```

然后添加数据库驱动依赖配置。

```xml
<dependency>
    <groupId>mysql</groupId>
    <artifactId>mysql-connector-java</artifactId>
</dependency>
```

最后需要加入 spring-boot-starter-data-jpa，它是 Spring Boot 项目访问数据库的核心依赖配置，加入此配置后，系统会自动导入 Spring Data 相关的核心数据访问接口包，以及 Hibernate 框架相关的依赖包，该配置如下。

```xml
<dependency>
    <groupId>org.springframework.boot</groupId>
    <artifactId>spring-boot-starter-data-jpa</artifactId>
</dependency>
```

修改后的完整 pom.xml 文件如下。

程序清单：codes/04/crudrepositorytest/pom.xml

```xml
<project xmlns="http://maven.apache.org/POM/4.0.0" xmlns:xsi="http://www.w3.org/2001/XMLSchema-instance"
    xsi:schemaLocation="http://maven.apache.org/POM/4.0.0 http://maven.apache.org/xsd/maven-4.0.0.xsd">
    <modelVersion>4.0.0</modelVersion>
    <groupId>org.fkit</groupId>
    <artifactId>crudrepositorytest</artifactId>
    <version>0.0.1-SNAPSHOT</version>
    <packaging>jar</packaging>
    <name>crudrepositorytest</name>
```

```xml
<url>http://maven.apache.org</url>
<parent>
    <groupId>org.springframework.boot</groupId>
    <artifactId>spring-boot-starter-parent</artifactId>
    <version>2.0.0.RELEASE</version>
</parent>

<properties>
    <project.build.sourceEncoding>UTF-8</project.build.sourceEncoding>
    <project.reporting.outputEncoding>UTF-8</project.reporting.outputEncoding>
    <java.version>1.8</java.version>
</properties>
<dependencies>
    <!--添加 spring-boot-starter-web 依赖-->
    <dependency>
        <groupId>org.springframework.boot</groupId>
        <artifactId>spring-boot-starter-web</artifactId>
    </dependency>
    <!-- 添加 spring-boot-starter-thymeleaf 模块依赖 -->
    <dependency>
        <groupId>org.springframework.boot</groupId>
        <artifactId>spring-boot-starter-thymeleaf</artifactId>
    </dependency>

    <!-- 添加 MySQL 依赖 -->
    <dependency>
        <groupId>mysql</groupId>
        <artifactId>mysql-connector-java</artifactId>
    </dependency>

    <!--添加 Spring Data JPA 依赖-->
    <dependency>
        <groupId>org.springframework.boot</groupId>
        <artifactId>spring-boot-starter-data-jpa</artifactId>
    </dependency>
</dependencies>
</project>
```

② 配置基本属性。

在 src/main/resources 包下新建一个全局配置文件，命名为 application.properties，在该配置文件中配置数据源和 JPA 相关的属性。

程序清单：codes/04/crudrepositorytest/src/main/resources/application.properties

```
###########################################################
### 数据源信息配置
###########################################################
# 数据库地址
spring.datasource.url = jdbc:mysql://localhost:3306/springdatajpa
# 用户名
spring.datasource.username = root
# 密码
spring.datasource.password =
# 数据库驱动
spring.datasource.driverClassName = com.mysql.jdbc.Driver
# 指定连接池中最大的活跃连接数
spring.datasource.max-active=20
# 指定连接池最大的空闲连接
spring.datasource.max-idle=8
# 指定必须保持连接的最小值
spring.datasource.min-idle=8
```

```
# 指定启动连接池时，初始建立的连接数量
spring.datasource.initial-size=10


############################################################
### JPA 持久化配置
############################################################
# 指定数据库的类型
spring.jpa.database = MySQL
# 指定是否需要在日志中显示 SQL 语句
spring.jpa.show-sql = true
# 指定自动创建|更新|验证数据库表结构等配置，配置成 update
# 表示如果数据库中存在持久化类对应的表就不创建，不存在就创建对应的表
spring.jpa.hibernate.ddl-auto = update
# 指定命名策略
spring.jpa.hibernate.naming.physical-strategy=org.hibernate.boot.model.naming.PhysicalNamingStrategyStandardImpl
# 指定数据库方言
spring.jpa.properties.hibernate.dialect = org.hibernate.dialect.MySQL5Dialect
```

③ 创建持久化类。

在 org.fkit.crudrepositorytest 包下新建 4 个包，分别是 bean（放置持久化类）、controller（控制器）、repository（定义数据访问接口的包）、service（业务逻辑处理类），在 bean 包中创建一个持久化类 User.java。

程序清单：codes/04/crudrepositorytest/src/main/java/org/fkit/crudrepositorytest/bean/User.java

```java
import javax.persistence.Entity;
import javax.persistence.GeneratedValue;
import javax.persistence.GenerationType;
import javax.persistence.Id;
import java.io.Serializable;

@Entity
// 用于标记持久化类，Spring Boot 项目加载后会自动根据持久化类建表
@Table(name="tb_user")
public class User  implements Serializable{

    /**
     * 使用@Id 指定主键。使用代码@GeneratedValue(strategy=GenerationType.IDENTITY)
     * 指定主键的生成策略，MySQL 默认为自增长
     *
     */
    @Id
    @GeneratedValue(strategy = GenerationType.IDENTITY)
    private int id;// 主键

    private String username;// 姓名, username

    private String loginName;

    private char sex;// 性别

    private int age; // 年龄

    // 省略构造器和 set/get 方法
}
```

④ 定义数据访问层接口。

在 org.fkit.crudrepositorytest.repository 包下新建一个接口，命名为 UserRepository，让该接口继承 CrudRepository 接口，以持久化对象 User 作为 CrudRepository 的第一个类型参数，表

示当前所操作的持久化对象类型，Integer 作为 CrudRepository 的第二个类型参数，用于指定 ID 类型，完整代码如下。

程序清单：codes/04/crudrepositorytest/src/main/java/org/fkit/crudrepositorytest/repository/UserRepository.java

```java
import org.fkit.crudrepositorytest.bean.User;
import org.springframework.data.repository.CrudRepository;

public interface UserRepository extends CrudRepository<User, Integer>{

}
```

在 Spring Boot 项目中，数据访问层无须提供实现，直接继承数据访问接口即可。

⑤ 定义业务层类。

程序清单：codes/04/crudrepositorytest/src/main/java/org/fkit/crudrepositorytest/service/UserService.java

```java
import javax.annotation.Resource;
import javax.transaction.Transactional;
import org.fkit.crudrepositorytest.bean.User;
import org.fkit.crudrepositorytest.repository.UserRepository;
import org.springframework.stereotype.Service;
@Service
public class UserService {

    // 注入UserRepository
    @Resource
    private UserRepository userRepository;

    /**
     * save、update、delete 方法需要绑定事务。使用@Transactional 进行事务的绑定
     *
     * 保存对象
     * @param User
     * @return 包含自动生成的 id 的 User 对象
     */
    @Transactional
    public User save(User User) {
        return userRepository.save(User);
    }

    /**
     * 根据 id 删除对象
     *
     * @param id
     */
    @Transactional
    public void delete(int id) {
        userRepository.deleteById(id);
    }

    /**
     * 查询所有数据
     *
     * @return 返回所有 User 对象
     */
    public Iterable<User> getAll() {
        return userRepository.findAll();
    }

    /**
     * 根据 id 查询数据
```

```java
 *
 * @return 返回id对应的User对象
 */
public User getById(Integer id) {
    // 根据id查询对应的持久化对象
    Optional<User>op = userRepository.findById(id);
    return op.get();
}

/**
 * 修改用户对象数据，持久化对象修改会自动更新到数据库
 *
 * @param user
 */
@Transactional
public void update(User user) {
    // 直接调用持久化对象的set方法修改对象的数据
    user.setUsername("孙悟空");
    user.setLoginName("swk");
}

}
```

在业务层中需要注入数据访问层对象，在上述代码中是通过@Resources 注解将 UserRepository 接口对应的实现类注入进来的。从这里可以看出，只要数据访问层接口实现了 CrudReposity，Spring Boot 项目会自动扫描该类并为该类创建实现类对象。@Transactional 注解用于声明方法的事物特性。

⑥ 定义控制器类。

在 org.fkit.crudrepositorytest.controller 包下新建一个控制器类，命名为 UserController，其代码如下。

程序清单：codes/04/crudrepositorytest/src/main/java/org/fkit/crudrepositorytest/controller/UserController.java

```java
import javax.annotation.Resource;
import org.fkit.crudrepositorytest.bean.User;
import org.fkit.crudrepositorytest.service.UserService;
import org.springframework.web.bind.annotation.RequestMapping;
import org.springframework.web.bind.annotation.RestController;

@RestController
@RequestMapping("/user")
public class UserController {

    // 注入UserService
    @Resource
    private UserService userService;

    @RequestMapping("/save")
    public String save() {
        User user = new User();
        user.setLoginName("dlei");
        user.setUsername("徐磊");
        user.setSex('男');
        user.setAge(3);
        userService.save(user);
        return "保存数据成功！";
    }

    @RequestMapping("/update")
```

```
    public String update() {
        // 修改的对象必须是持久化对象,所以先从数据库查询 id 为 1 的对象开始修改
        User user = userService.getById(1);
        userService.update(user);
        return "修改成功!";
    }

    @RequestMapping("/delete")
    public String delete() {
        userService.delete(1);
        return "删除数据成功!";
    }

    @RequestMapping("/getAll")
    public Iterable<User> getAll() {
        // 查询所有的用户数据
        return userService.getAll();
    }
}
```

⑦ 测试应用。

启动 MySQL 数据库,在数据库中创建名为 springdatajpa 的数据库,执行如下脚本。

```
CREATE DATABASE springdatajpa;
```

然后在 org.fkit.crudrepositorytest 包下新建 App.java 启动类,App.java 和之前的项目一致,此处不再赘述。右击该类运行 main 方法。Spring Boot 项目启动后,JPA 会在数据库中自动创建持久化类对应的 tb_user 表。需要注意的是,按照 JPA 规范,实体类 User 的属性 loginName 映射到数据库的时候,自动生成的数据库列名是 login_name。

在浏览器中输入 URL 来测试应用。

```
http://127.0.0.1:8080/user/save
```

请求会提交到 UserController 类的 save 方法进行处理,如果该方法保存用户成功即可返回字符串"保存数据成功!",如图 4.5 所示。

图 4.5 保存数据

查看数据库的表信息,如图 4.6 所示。

图 4.6 保存数据后的表信息

测试修改用户,在浏览器中输入如下地址。

```
http://127.0.0.1:8080/user/update
```

请求会提交到 UserController 类的 update 方法进行处理,如果该方法修改用户成功即可返回字符串"修改数据成功!",如图 4.7 所示。

图 4.7 修改用户

查看数据库的表信息，如图 4.8 所示。

图 4.8　修改数据后的表信息

测试查询所有用户数据，在浏览器中输入如下地址。

`http://127.0.0.1:8080/user/getAll`

请求会提交到 UserController 类的 getAll 方法进行处理，如果该方法执行查询所有用户成功，即可返回查询到的所有用户信息的 JSON 格式的字符串，如图 4.9 所示。

图 4.9　查询所有用户数据

测试删除用户数据，在浏览器中输入如下地址。

`http://127.0.0.1:8080/user/delete`

请求会提交到 UserController 类的 delete 方法进行处理，如图 4.10 所示。

图 4.10　删除用户信息

查看数据库的表信息，如图 4.11 所示。

图 4.11　删除数据后的表信息

PagingAndSortingRepository

PagingAndSortingRepository 继承自 CrudRepository 接口，除了拥有 CrudRepository 接口的功能之外，PagingAndSortingRepository 接口还提供了分页与排序功能。PagingAndSortingRepository 接口的源码如下。

程序清单：org/springframework/data/repository/PagingAndSortingRepository.java

```java
package org.springframework.data.repository;
import java.io.Serializable;
import org.springframework.data.domain.Page;
import org.springframework.data.domain.Pageable;
import org.springframework.data.domain.Sort;

@NoRepositoryBean
public interface PagingAndSortingRepository<T, ID extends Serializable> extends CrudRepository<T, ID> {
    Iterable<T> findAll(Sort sort);
    Page<T> findAll(Pageable pageable);
}
```

该接口提供的排序和分页方法如下。
- Iterable<T> findAll(Sort sort)：按照指定的排序对象规则查询出实体对象数据。
- Page<T> findAll(Pageable pageable)：分页查询实体对象，包含排序功能操作。

示例：PagingAndSortingRepository 接口访问数据

PagingAndSortingRepository 继承自 CrudRepository 接口，所以除了拥有 CrudRepository 的功能之外，它还增加了排序和分页查询的功能。现创建一个新的 Maven 项目，命名为 pagingandsortingrepositorytest。按照 Maven 项目的规范，在 src/main/下新建一个名为 resources 的文件夹。

① 修改 pom.xml 文件。

在 pom.xml 文件中加入 mysql-connector-java 依赖以及 spring-boot-starter-data-jpa 依赖，pom.xml 文件配置与"示例：CrudRepository 接口访问数据"相同。

程序清单：codes/04/pagingandsortingrepositorytest/pom.xml

```xml
<project xmlns="http://maven.apache.org/POM/4.0.0" xmlns:xsi="http://www.w3.org/2001/XMLSchema-instance"
    xsi:schemaLocation="http://maven.apache.org/POM/4.0.0 http://maven.apache.org/xsd/maven-4.0.0.xsd">
    <modelVersion>4.0.0</modelVersion>
    <groupId>org.fkit</groupId>
    <artifactId>pagingandsortingrepositorytest</artifactId>
    <version>0.0.1-SNAPSHOT</version>
    <packaging>jar</packaging>
    <name>pagingandsortingrepositorytest</name>
    <url>http://maven.apache.org</url>
    <parent>
        <groupId>org.springframework.boot</groupId>
        <artifactId>spring-boot-starter-parent</artifactId>
        <version>2.0.0.RELEASE</version>
    </parent>

    <properties>
        <project.build.sourceEncoding>UTF-8</project.build.sourceEncoding>
        <project.reporting.outputEncoding>UTF-8</project.reporting.outputEncoding>
        <java.version>1.8</java.version>
    </properties>
    <dependencies>
        <!--添加 spring-boot-starter-web 依赖... -->
        <dependency>
            <groupId>org.springframework.boot</groupId>
            <artifactId>spring-boot-starter-web</artifactId>
        </dependency>
        <!-- 添加 spring-boot-starter-thymeleaf 模块依赖 -->
        <dependency>
             <groupId>org.springframework.boot</groupId>
             <artifactId>spring-boot-starter-thymeleaf</artifactId>
        </dependency>

        <!-- 添加 MySQL 依赖 -->
        <dependency>
            <groupId>mysql</groupId>
            <artifactId>mysql-connector-java</artifactId>
        </dependency>
        <!-- 添加 Spring Data JPA 依赖 -->
        <dependency>
            <groupId>org.springframework.boot</groupId>
            <artifactId>spring-boot-starter-data-jpa</artifactId>
```

```xml
            </dependency>
        </dependencies>
</project>
```

② 配置基本属性。

在 src/main/resources 包下新建一个全局配置文件，命名为 application.properties，在该配置文件中配置数据源和 JPA 相关的属性。

程序清单：codes/04/pagingandsortingrepositorytest/src/main/resources/application.properties

```
###########################################################
### 数据源信息配置
###########################################################
# 数据库地址
spring.datasource.url = jdbc:mysql://localhost:3306/springdatajpa
# 用户名
spring.datasource.username = root
# 密码
spring.datasource.password =
# 数据库驱动
spring.datasource.driverClassName = com.mysql.jdbc.Driver
# 指定连接池中最大的活跃连接数
spring.datasource.max-active=20
# 指定连接池最大的空闲连接数
spring.datasource.max-idle=8
# 指定必须保持连接的最小值
spring.datasource.min-idle=8
# 指定启动连接池时，初始建立的连接数量
spring.datasource.initial-size=10


###########################################################
### JPA 持久化配置
###########################################################
# 指定数据库的类型
spring.jpa.database = MySQL
# 指定是否需要在日志中显示 SQL 语句
spring.jpa.show-sql = true
# 指定自动创建|更新|验证数据库表结构等配置，配置成 update
# 表示如果数据库中存在持久化类对应的表就不创建，不存在就创建对应的表
spring.jpa.hibernate.ddl-auto = update
# 指定命名策略
spring.jpa.hibernate.naming-strategy = org.hibernate.cfg.ImprovedNamingStrategy
# 指定数据库方言
spring.jpa.properties.hibernate.dialect = org.hibernate.dialect.MySQL5Dialect
```

③ 创建持久化类。

在 org.fkit.pagingandsortingrepositorytest 包下新建 4 个包，分别是 bean（放置持久化类）、controller（控制器）、repository（定义数据访问接口的包）、service（业务逻辑处理类），在 bean 包中创建一个持久化类 Article.java，其代码如下。

程序清单：codes/04/pagingandsortingrepositorytest/src/main/java/org/fkit/pagingandsortingrepositorytest/bean/Article.java

```java
import java.util.Date;
import javax.persistence.Entity;
import javax.persistence.GeneratedValue;
import javax.persistence.GenerationType;
import javax.persistence.Id;
import javax.persistence.Table;
/**
```

```
 * 商品对象
 * @author   xulei
 * */
@Entity
@Table(name="tb_article")
public class Article implements Serializable {
    @Id
    @GeneratedValue(strategy = GenerationType.IDENTITY)
    private int id;
    private String title;
    private String supplier;
    private Double price;
    private String locality;
    private Date putawayDate;
    private int storage;
    private String image;
    private String description;
    private Date createDate;

// 省略构造器和set/get方法……
}
```

④ 定义数据访问层接口。

在 org.fkit.pagingandsortingrepositorytest.repository 包下新建一个接口,命名为 ArticleRepository,让该接口继承 PagingAndSortingRepository 接口,以持久化对象 Article 作为 PagingAndSortingRepository 的第一个类型参数,表示当前所操作的持久化对象类型,Integer 作为 PagingAndSortingRepository 的第二个类型参数,用于指定 ID 类型,完整代码如下。

> 程序清单:codes/04/pagingandsortingrepositorytest/src/main/java/org/fkit/
> pagingandsortingrepositorytest/repository/ArticleRepository.java

```
import org.fkit.pagingandsortingrepositorytest.bean.Article;
import org.springframework.data.repository.PagingAndSortingRepository;

public interface ArticleRepository extends PagingAndSortingRepository<Article, Integer> {

}
```

在 Spring Boot 项目中,数据访问层无须提供实现,直接继承数据访问接口即可。

⑤ 定义业务层类。

> 程序清单:codes/04/pagingandsortingrepositorytest/src/main/java/org/
> fkit/pagingandsortingrepositorytest/service/ArticleService.java

```
import javax.annotation.Resource;
import org.fkit.pagingandsortingrepositorytest.bean.Article;
import org.fkit.pagingandsortingrepositorytest.repository.ArticleRepository;
import org.springframework.data.domain.Page;
import org.springframework.data.domain.Pageable;
import org.springframework.data.domain.Sort;
import org.springframework.stereotype.Service;

@Service
public class ArticleService {

    // 数据访问层接口对象
    @Resource
    private ArticleRepository articleRepository;

    public Iterable<Article> findAllSort(Sort sort) {
        return articleRepository.findAll(sort);
    }
```

```java
    public Page<Article> findAll(Pageable page) {
        return articleRepository.findAll(page);
    }
}
```

在业务层中需要注入数据访问层对象,在上述代码中是通过@Resources注解将ArticleRepository接口对应的实现类注入进来的。

⑥ 定义控制器类。

在 org.fkit.pagingandsortingrepositorytest.controller 包中先新建一个控制器类,命名为ArticleController,其代码如下。

程序清单:codes/04/pagingandsortingrepositorytest/src/main/java/org/fkit/pagingandsortingrepositorytest/controller/ArticleController.java

```java
import java.util.List;
import javax.annotation.Resource;
import org.fkit.pagingandsortingrepositorytest.bean.Article;
import org.fkit.pagingandsortingrepositorytest.service.ArticleService;
import org.springframework.data.domain.Page;
import org.springframework.data.domain.PageRequest;
import org.springframework.data.domain.Pageable;
import org.springframework.data.domain.Sort;
import org.springframework.web.bind.annotation.RequestMapping;
import org.springframework.web.bind.annotation.RestController;

@RestController
@RequestMapping("/article")
public class ArticleController {

    // 注入ArticleService
    @Resource
    private ArticleService articleService;

    @RequestMapping("/sort")
    public Iterable<Article> sortArticle() {
        // 指定排序参数对象:根据id,进行降序查询
        Sort sort = new Sort(Sort.Direction.DESC, "id");
        Iterable<Article> articleDatas = articleService.findAllSort(sort);
        return articleDatas;
    }

    @RequestMapping("/pager")
    public List<Article> sortPagerArticle(int pageIndex) {
        // 指定排序参数对象:根据id,进行降序查询
        Sort sort = new Sort(Sort.Direction.DESC, "id");
        /**
         * 封装分页实体
         * 参数1:pageIndex 表示当前查询的是第几页(默认从0开始,0表示第1页)
         * 参数2:表示每页展示多少数据,现在设置每页展示2条数据
         * 参数3:封装排序对象,根据该对象的参数指定根据id降序查询
         */
        Pageable page = PageRequest.of(pageIndex - 1, 2, sort);
        Page<Article> articleDatas = articleService.findAll(page);
        System.out.println("查询总页数:" + articleDatas.getTotalPages());
        System.out.println("查询总记录数:" + articleDatas.getTotalElements());
        System.out.println("查询当前第几页:" + articleDatas.getNumber() + 1);
        System.out.println("查询当前页面的记录数:" + articleDatas.getNumberOfElements());
        // 查询出的结果数据集合
        List<Article> articles = articleDatas.getContent();
        System.out.println("查询当前页面的集合:" + articles);
```

```
        return articles;
    }
}
```

⑦ 测试应用。

启动 MySQL 数据库，继续使用之前在数据库中创建的 springdatajpa 数据库，然后在 org.fkit.pagingandsortingrepositorytest 包下新建 App.java 启动类，App.java 和之前的项目一致，此处不再赘述。右击该类运行 main 方法。Spring Boot 项目启动后，JPA 会在数据库中自动创建持久化类对应的 tb_article 表。然后再打开数据库管理工具执行如下测试数据脚本。

程序清单： codes/04/pagingandsortingrepositorytest/scripts/db.sql

```sql
USE springdatajpa;
INSERT INTO tb_article (
  TITLE,
  SUPPLIER,
  PRICE,
  LOCALITY,
  PUTAWAY_DATE,
  STORAGE,
  IMAGE,
  DESCRIPTION,
  CREATE_DATE
)
VALUES
  (
    '疯狂Java讲义(附光盘)。',
    '李刚 著',
    108.8,
    '电子工业出版社',
    NULL,
    100,
    'java.jpg',
    '疯狂源自梦想，技术成就辉煌 本书来自作者3年的Java培训经历，凝结了作者近3000小时的授课经验，总结了几百个Java学员学习过程中的典型错误。',
    '2008-10-01'
  ),
  (
    '轻量级Java EE企业应用实战。',
    '李刚 著',
    79.8,
    '电子工业出版社',
    NULL,
    100,
    'ee.jpg',
    '本书主要介绍以Spring+Hibernate为基础的Java EE应用。',
    '2008-11-01'
  ),
  (
    'Spring+MyBatis企业应用实战。',
    '肖文吉 著',
    58,
    '电子工业出版社',
    NULL,
    100,
    'Spring+MyBatis.jpg',
    '媲美于SSH组合的轻量级Java EE开发方式。',
    '2017-01-01'
  ),
  (
    '疯狂Ajax讲义(含CD光盘1张)。',
```

'李刚 著',
66.6,
'电子工业出版社',
NULL,
100,
'ajax.jpg',
'异步访问技术，很多网站都在用。',
'2011-07-01'
),
(
'疯狂iOS讲义（基础篇）（含光盘1张）',
'李刚 著',
85.6,
'电子工业出版社',
NULL,
100,
'ios.jpg',
'基于iOS全新版本彻底升级，Swift和Objective-C双语讲解。',
'2016-05-01'
),
(
'魔戒：插图珍藏版（200套限量编号版随机发送！）。奴役全世界的力量。',
'(英国) J.R.R.托尔金 著',
132.3,
'译林出版社',
NULL,
100,
'22566493-1_b.jpg',
'天真无邪的哈比男孩佛罗多继承了一枚戒，这是一次异常艰险的远征……',
'2012-06-18'
),
(
'肖申克的救赎。',
'(美) 金 著,施寄青,赵永芬,齐若兰',
19.9,
'人民文学出版社',
NULL,
100,
'9198692-1_l.jpg',
'本书是斯蒂芬•金最为人津津乐道的杰出代表作。',
'2012-06-11'
),
(
'权威定本四大名著：西游记 水浒传 三国演义 红楼梦 全国独家',
'[明] 吴承恩,[明] 施耐庵,[明] 罗贯中,[清] 曹雪芹 著；黄肃秋 注',
145.8,
'清华大学出版社',
NULL,
100,
'20605371-1_a.jpg',
'权威定本四大名著：红楼梦 三国演义 水浒传 西游记。',
'2012-06-11'
),
(
'万物生光辉',
'[英] 吉米•哈利 著；余国芳,谢瑶玲 译',
23.6,
'湖北教育出版社',
NULL,
100,
'22639083-1_a.jpg',

```
    '邂逅最可爱的动物，感受最纯真的幽默。',
    '2012-06-11'
),
(
    '爱你是最好的时光',
    '匪我思存 著 ',
    17.6,
    '译林出版社',
    NULL,
    100,
    '22630101-1_a.jpg',
    '终极大结局完结篇。',
    '2012-06-11'
),
(
    '白鹿原',
    '陈忠实 著',
    360,
    '春风文艺出版社',
    NULL,
    100,
    '22541642-1_a.jpg',
    '中国首部当代名家名篇宣纸线装书,陈忠实先生亲笔签名签章限量珍藏版。',
    '2012-06-11'
),
(
    '疯狂 XML 讲义(含光盘 1 张),包括 DTD、Schema 等技术的深入讲解。',
    '李刚 著',
    48.8,
    '电子工业出版社',
    NULL,
    100,
    'xml.jpg',
    '本书主要以 XML 为核心,深入介绍了 XML 的各种相关知识。',
    '2012-06-11'
),
(
    ' Struts 2.x 权威指南（第 3 版）(含 DVD 光盘 1 张)',
    '李刚 著',
    79.20,
    '电子工业出版社',
    NULL,
    100,
    'struts.jpg',
    '全面介绍了 Struts 2 框架的各知识点',
    '2012-06-11'
),
(
    '疯狂 Android 讲义(含 CD 光盘 1 张)',
    '李刚 著',
    60.6,
    '电子工业出版社',
    NULL,
    100,
    'android.jpg',
    '本书全面地介绍了 Android 应用开发的相关知识。',
    '2012-06-11'
);
```

测试脚本执行成功以后，打开浏览器，输入如下 URL：

http://127.0.0.1:8080/article/sort

请求会提交到 ArticleController 类的 sortArticle 方法进行处理，该方法执行排序查询，将 tb_article 表中的数据以降序的方式查询出来并以 JSON 格式返回浏览器，如图 4.12 所示。

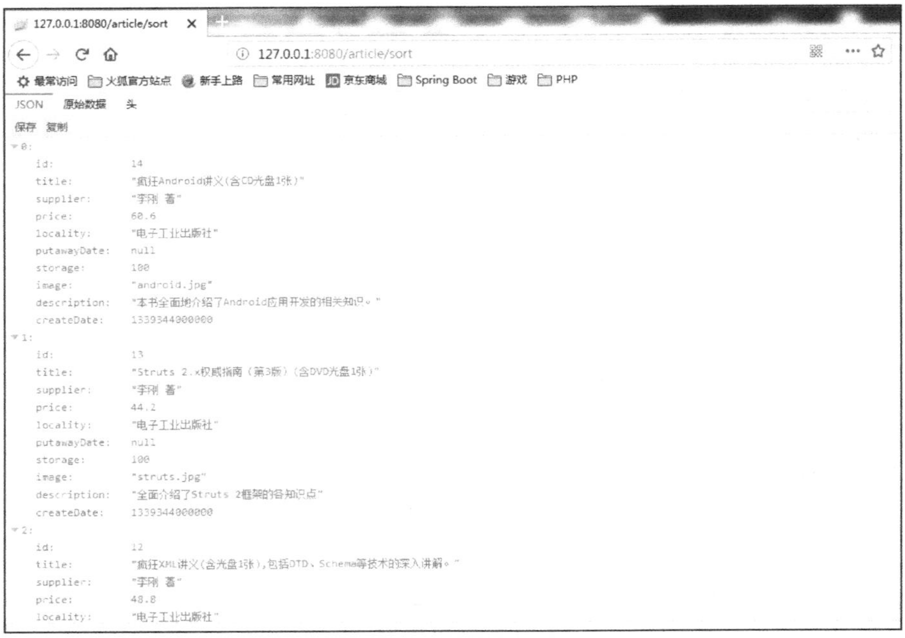

图 4.12　排序查询商品数据

测试分页查询第一页商品数据，在浏览器中输入如下地址。

```
http://127.0.0.1:8080/article/pager?pageIndex=1
```

请求会提交到 ArticleController 类的 sortPagerArticle 方法进行处理，该方法接收 pageIndex 参数，根据该参数确定查询第几页数据，查询到第 1 页数据，如图 4.13 所示。

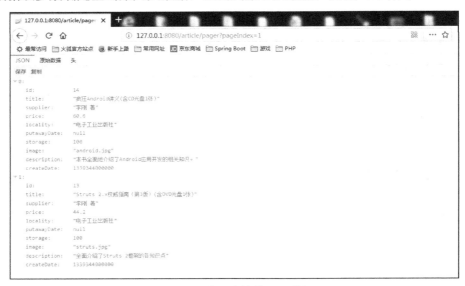

图 4.13　分页查询第 1 页数据

测试分页查询第 1 页商品数据，在浏览器中输入如下地址。

```
http://127.0.0.1:8080/article/pager?pageIndex=2
```

请求会提交给 ArticleController 类的 sortPagerArticle 方法进行处理，该方法接收 pageIndex 参数，根据该参数确定查询第几页数据，查询到第 2 页数据，如图 4.14 所示。

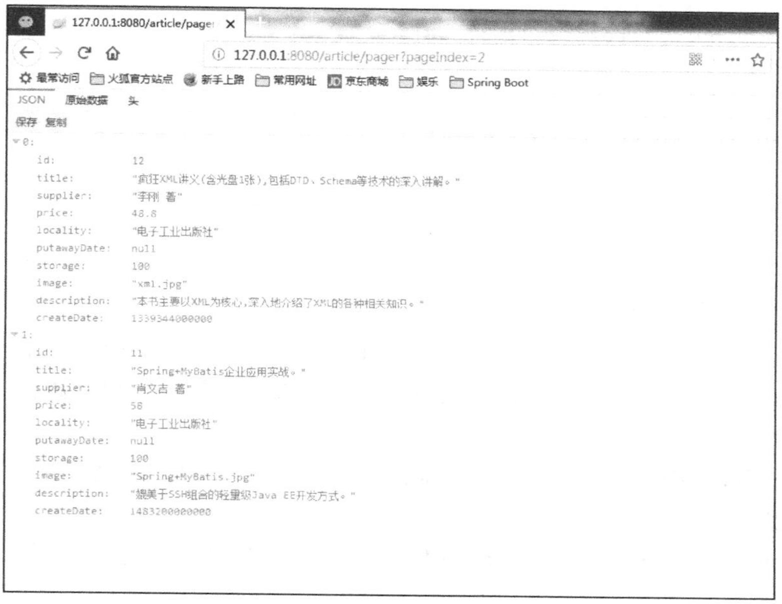

图 4.14　分页查询第 2 页数据

4.2.2　Spring Data JPA 开发

Spring Data JPA 是 Spring Data 技术下的子项目，使用 Spring Data JPA 访问数据只需数据访问层接口实现 JpaRepository 接口即可，Spring Data JPA 项目的数据访问层接口可以在不写实现的情况下实现大部分的数据访问操作，使用起来十分方便。由于 JpaRepository 接口继承了 PagingAndSortingRepository 接口，所以 JpaRepository 接口也拥有了 PagingAndSortingRepository 接口的相关功能操作，JpaRepository 接口封装了常用的一些方法，使用方式都类似，同时按照 Spring Data 定义的规则，也可以在数据访问层接口中自定义一些查询方法来完成数据的访问操作。

JpaRepository

JpaRepository 继承自 PagingAndSortingRepository 接口，因此将获得更多的操作功能，JpaRepository 是基于 JPA 的 Repository 接口，它极大地减少了 JPA 作为数据访问的代码，JpaRepository 是实现 Spring Data JPA 技术访问数据库的关键接口。JpaRepository 接口的源码如下。

程序清单：org/springframework/data/jpa/repository/JpaRepository.java

```
package org.springframework.data.jpa.repository;
import java.io.Serializable;
import java.util.List;
import javax.persistence.EntityManager;
import org.springframework.data.domain.Example;
import org.springframework.data.domain.Sort;
import org.springframework.data.repository.NoRepositoryBean;
import org.springframework.data.repository.PagingAndSortingRepository;
import org.springframework.data.repository.query.QueryByExampleExecutor;

@NoRepositoryBean
```

```
public interface JpaRepository<T, ID extends Serializable>
    extends PagingAndSortingRepository<T, ID>, QueryByExampleExecutor<T> {
    List<T> findAll();
    List<T> findAll(Sort sort);
    List<T> findAll(Iterable<ID> ids);
    <S extends T> List<S> save(Iterable<S> entities);
    void flush();
    <S extends T> S saveAndFlush(S entity);
    void deleteInBatch(Iterable<T> entities);
    void deleteAllInBatch();
    T getOne(ID id);
    <S extends T> List<S> findAll(Example<S> example);
    <S extends T> List<S> findAll(Example<S> example, Sort sort);
}
```

该接口提供的常用方法如下。

> List<T> findAll()：查询所有的实体对象数据，返回的是一个 List 集合。
> List<T> findAll(Sort sort)：按照指定的排序对象规则查询实体对象数据，返回的是一个 List 集合。
> List<T> findAll(Iterable<ID> ids)：根据所提供的实体对象 id，将对应的实体全部查询出来，返回的是一个 List 集合。
> <S extends T> List<S> save(Iterable<S> entities)：将提供的集合中的实体对象数据保存到数据库。
> void flush()：将缓存的对象数据操作更新到数据库中。
> <S extends T> S saveAndFlush(S entity)：保存对象的同时立即更新到数据库。
> void deleteInBatch(Iterable<T> entities)：批量删除提供的实体对象。
> void deleteAllInBatch()：批量删除所有的实体对象数据。
> T getOne(ID id)：根据 id 查询出对应的实体对象。
> <S extends T> List<S> findAll(Example<S> example)：根据提供的 example 实例查询实体对象数据。
> <S extends T> List<S> findAll(Example<S> example, Sort sort)：根据提供的 example 实例查询实体对象数据，同时支持排序查询。

示例：简单条件查询

按照 Spring Data 的规则，我们可以通过定义在 Repository 接口下的方法名称来执行查询等操作，查询的方法名称必须以 find、get、read 开头，同时，涉及条件查询时，Spring Data JPA 支持将条件属性定义在数据访问层接口下的方法名中，条件属性通过条件关键字连接。需要注意的是：条件属性的首字母必须大写，接下来以示例的方式进行讲解。

① 创建一个新的 Maven 项目，命名为 simplespringdatajpatest，修改 pom.xml 文件，参考前一节示例配置，此处不再赘述。

② 配置基本属性。

在 src/main/resources 包下新建一个全局配置文件，命名为 application.properties，在该配置文件中配置数据源和 JPA 相关的属性，内容请参考前一节的示例配置，此处不再赘述。

③ 创建持久化类。

在 org.fkit.simplespringdatajpatest 包下新建 4 个包，分别是 bean（放置持久化类）、controller

（控制器）、repository（定义数据访问接口的包）、service（业务逻辑处理类），在 bean 包中创建一个持久化类 Student.java，其代码如下。

程序清单：codes/04/simplespringdatajpatest/src/main/java/org/fkit/
simplespringdatajpatest/bean/Student.java

```java
import javax.persistence.Entity;
import javax.persistence.GeneratedValue;
import javax.persistence.GenerationType;
import javax.persistence.Id;

@Entity
@Table(name="tb_student")
public class Student  implements Serializable{
    @Id
    @GeneratedValue(strategy = GenerationType.IDENTITY)
    private int id;
    private String name;
    private String address;
    private int age;
    private char sex;
    // 省略构造器和 set/get 方法……
}
```

④ 定义数据访问层接口。

在 org.fkit.simplespringdatajpatest.repository 包下新建一个接口，命名为 StudentRepository，让该接口继承 JpaRepository 接口，以持久化对象 Student 作为 JpaRepository 的第一个类型参数，表示当前所操作的持久化对象类型，Integer 作为 JpaRepository 的第二个类型参数，用于指定 ID 类型，完整代码如下。

程序清单：codes/04/simplespringdatajpatest/src/main/java/org/fkit/
simplespringdatajpatest/repository/StudentRepository.java

```java
import java.util.List;
import org.fkit.simplespringdatajpatest.bean.Student;
import org.springframework.data.jpa.repository.JpaRepository;

public interface StudentRepository extends JpaRepository<Student, Integer> {
    /**
     * 通过学生姓名来查询学生对象
     * 此方法相当于 JPQL 语句代码:select s from Student s where s.name = ?1
     * @param name 参数
     * @return Student 对象
     */
    Student findByName(String name);

    /**
     * 通过名字和地址查询学生信息
     * 此方法相当于 JPQL 语句代码:select s from Student s where s.name = ?1 and s.address=?2
     * @param name
     * @param address
     * @return 包含 Student 对象的 List 集合
     */
    List<Student> findByNameAndAddress(String name , String address);

    /**
     * 通过学生姓名模糊查询学生信息
     * 此方法相当于 JPQL 语句代码:select s from Student s where s.name like ?1
     * @param name 参数
     * @return 包含 Student 对象的 List 集合
```

```
    */
    List<Student> findByNameLike(String name);
}
```

- 从上述代码可以看出，findByName(String name)方法等同于 JPQL 语句：select s from Student s where s.name = ?1，因为 by 是条件关键字，而 name 是条件属性，条件关键字和条件属性的首字母必须大写，所以上述方法名的定义即 findByName。从这里可以看出，在 Spring Data JPA 中可以直接通过在数据访问层定义方法名称即可进行数据的访问操作了。
- findByNameAndAddress(String name,String address)方法相当于 JPQL 语句：select s from Student s where s.name = ?1 and s.address=?2。其中 by 和 and 是条件关键字，name 和 address 是条件属性。
- findByNameLike(String name)方法，相当于 JPQL 语句：select s from Student s where s.name like ?1。其中 by 和 like 是条件关键字，name 是条件属性。

除此之外，还有很多形式的命名查询方式，如表 4-1 所示。

表 4-1 查询关键字

关键字	例子	SQL 说明
And	findByNameAndAddress	Where s.name=?1 And s.address=?2
Or	findByNameOrAddress	Where s.name=?1 Or s.address=?2
Between	findByAgeBetween	Where s.age between ?1 and ?2
LessThan	findByAgeLessThan	Where s.age < ?1
LessThanEqual	findByAgeLessThanEqual	Where s.age <= ?1
GreaterThan	findByAgeGreaterThan	Where s.age > ?1
GreaterThanEqual	findByAgeGreaterThanEqual	Where s.age >= ?1
After	findByStartDateAfter	where s.startDate > ?1
Before	findByStartDateBefore	where s.startDate < ?1
IsNull	findByAgeIsNull	Where s.age is null
IsNotNull，NotNull	findByAge(Is)NotNull	Where s.age is not null
Like	findByNameLike	Where s.name like ?1
NotLike	findByNameNotLike	Where s.name not like ?1
StartingWith	findByNameStartingWith	Where s.name like ?1(以 xxx 参数开头的模糊查询，参数最终形式是 xxx%)
EndingWith	findByNameEndingWith	Where s.name like ?1(以 xxx 参数结尾的模糊查询，参数最终形式是%xxx)
Containing	findByNameContaining	Where s.name like ?1(包含 xxx 参数的模糊查询，参数最终形式是%xxx%)
OrderBy	findBySexOrderByAgeDesc	Where s.sex=?1 order by s.age desc
Not	findByNameNot	Where s.name <> ?1
In	findByAgeIn(Collection<age> ages)	Where s.age in ?1
NotIn	findByAgeNotIn(Collection<age> ages)	Where s.age not in ?1
True	findByVipTrue	Where s.vip=true
False	findByVipFalse	Where s.vip=false

⑤ 定义业务层类。

程序清单：codes/04/simplespringdatajpatest/src/main/java/org/fkit/
simplespringdatajpatest/service/StudentService.java

```
import java.util.List;
import javax.annotation.Resource;
```

```java
import org.fkit.simplespringdatajpatest.bean.Student;
import org.fkit.simplespringdatajpatest.repository.StudentRepository;
import org.springframework.stereotype.Service;
import org.springframework.transaction.annotation.Transactional;

@Service
public class StudentService {

    // 注入数据访问层接口对象
    @Resource
    private StudentRepository studentRepository;

    @Transactional
    public void saveAll(List<Student> students) {
        studentRepository.saveAll(students);
    }

    public Student getStuByName(String name) {
        return studentRepository.findByName(name);
    }

    public List<Student> getStusByNameAndAddr(String name, String address) {
        return studentRepository.findByNameAndAddress(name, address);
    }

    public List<Student> getStusByNameLike(String name) {
        return studentRepository.findByNameLike("%"+name+"%");
    }
}
```

在业务层中需要注入数据访问层对象，在上述代码中是通过@Resources 注解将 StudentRepositor 接口对应的实现类对象注入进来的。

⑥ 定义控制器类。

在 org.fkit.simplespringdatajpatest.controller 包中新建一个控制器类,命名为 StudentController,完整代码如下。

程序清单：codes/04/simplespringdatajpatest/src/main/java/org/fkit/simplespringdatajpatest/controller/StudentController.java

```java
import java.util.ArrayList;
import java.util.List;
import javax.annotation.Resource;
import org.fkit.simplespringdatajpatest.bean.Student;
import org.fkit.simplespringdatajpatest.service.StudentService;
import org.springframework.web.bind.annotation.RequestMapping;
import org.springframework.web.bind.annotation.RestController;

@RestController
@RequestMapping("/student")
public class StudentController {

    // 注入 StudentService
    @Resource
    private StudentService studentService;

    @RequestMapping("/save")
    public String save() {
        Student swk = new Student();
        swk.setAddress("广州");
        swk.setName("孙悟空");
        swk.setAge(700);
        swk.setSex('男');
```

```
        Student zzj = new Student();
        zzj.setAddress("广州");
        zzj.setName("蜘蛛精");
        zzj.setAge(700);
        zzj.setSex('女');

        Student nmw = new Student();
        nmw.setAddress("广州");
        nmw.setName("牛魔王");
        nmw.setAge(500);
        nmw.setSex('男');

        List<Student> students = new ArrayList<>();
        students.add(swk);
        students.add(zzj);
        students.add(nmw);

        studentService.saveAll(students);
        return "保存学生对象成功";
    }

    @RequestMapping("/name")
    public Student getByName(String name) {
        return studentService.getStuByName(name);
    }

    @RequestMapping("/nameAndAddress")
    public List<Student> getByNameAndAddress(String name, String address) {
        return studentService.getStusByNameAndAddr(name, address);
    }

    @RequestMapping("/nameLike")
    public List<Student> getByNameLile(String name) {
        return studentService.getStusByNameLike(name);
    }

}
```

⑦ 测试应用。

启动 MySQL 数据库，继续使用之前在数据库中创建的 springdatajpa 数据库，然后在 org.fkit.simplespringdatajpatest 包下新建 App.java 启动类，App.java 和之前的项目一致，此处不再赘述。右击该类运行 main 方法。Spring Boot 项目启动后，JPA 会在数据库中自动创建持久化类对应的 tb_student 表。

测试添加学生信息，在浏览器中输入如下地址。

http://127.0.0.1:8080/student/save

请求会提交到 StudentController 类的 save 方法进行处理，执行完成返回"保存学生对象成功"，如图 4.15 所示。

图 4.15 保存学生对象数据

测试根据姓名查询学生对象，在浏览器中输入如下地址。

http://127.0.0.1:8080/student/name?name=孙悟空

请求会提交到 StudentController 类的 getByName 方法进行处理,执行完成返回查询出的学生对象,如图4.16所示。

图 4.16　根据姓名查询学生对象数据

测试根据姓名和地址查询学生对象,在浏览器中输入如下地址。

http://127.0.0.1:8080/student/nameAndAddress?name=蜘蛛精&address=广州

请求会提交到 StudentController 类的 getByNameAndAddr 方法进行处理,执行完成返回查询出的学生对象,如图4.17所示。

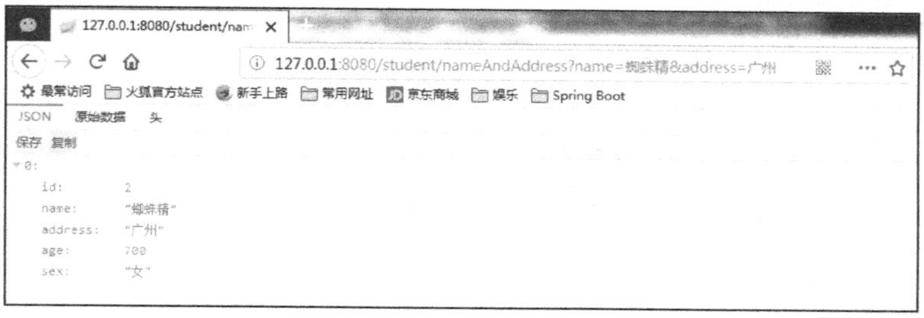

图 4.17　根据姓名和地址查询学生对象数据

测试根据姓名模糊查询学生对象,在浏览器中输入如下地址:

http://127.0.0.1:8080/student/nameLike?name=魔王

请求会提交到 StudentController 类的 getByNameLike 方法进行处理,执行完成返回查询出的学生对象,如图4.18所示。

图 4.18　根据姓名模糊查询学生对象数据

示例:关联查询和@Query查询

按照 Spring Data 的规则,对于两个有关联关系的对象的查询,可以通过方法名中的"_"下画线来标识,同时 Spring Data JPA 还支持用@Query 注解定义在数据访问层接口的方法上实现查询。接下来以示例的方式进行讲解。

① 创建一个新的 Maven 项目,命名为 springdatajpaquerytest,修改 pom.xml 文件,细节参考"简单条件查询"示例配置,内容完全一样,此处不再赘述。

② 配置基本属性。

在 src/main/resources 包下新建一个全局配置文件，命名为 application.properties，同时在 MySQL 数据库中执行如下命令：

```
CREATE DATABASE springdatajpaquery;  #创建一个新的数据库，名称是 springdatajpaquery
```

在 application.properties 配置文件中配置数据源和 jpa 相关的属性，详细代码如下。

```
############################################################
### 数据源信息配置
############################################################
# 数据库地址
spring.datasource.url = jdbc:mysql://localhost:3306/springdatajpaquery
# 用户名
spring.datasource.username = root
# 密码
spring.datasource.password =
# 数据库驱动
spring.datasource.driverClassName = com.mysql.jdbc.Driver
# 指定连接池中最大的活跃连接数
spring.datasource.max-active=20
# 指定连接池最大的空闲连接数
spring.datasource.max-idle=8
# 指定必须保持连接的最小值
spring.datasource.min-idle=8
# 指定启动连接池时，初始建立的连接数量
spring.datasource.initial-size=10
############################################################
### JPA 持久化配置
############################################################
# 指定数据库的类型
spring.jpa.database = MySQL
# 指定是否需要在日志中显示 SQL 语句
spring.jpa.show-sql = true
# 指定自动创建|更新|验证数据库表结构等配置，配置成 update
# 表示如果数据库中存在持久化类对应的表就不创建，不存在就创建对应的表
spring.jpa.hibernate.ddl-auto = update
# 指定命名策略
spring.jpa.hibernate.naming-strategy = org.hibernate.cfg.ImprovedNamingStrategy
# 指定数据库方言
spring.jpa.properties.hibernate.dialect = org.hibernate.dialect.MySQL5Dialect
```

③ 创建持久化类。

在 org.fkit.springdatajpaquerytest.repository 包下新建 4 个包，分别是 bean（放置持久化类的）、controller（控制器）、repository（定义数据访问接口的包）、service（业务逻辑处理类），在 bean 包创建两个持久化类 Student.java 以及 Clazz.java，完整代码如下。

程序清单：codes/04/springdatajpaquerytest/src/main/java/org/fkit/springdatajpaquerytest/bean/Student.java

```java
import javax.persistence.Entity;
import javax.persistence.FetchType;
import javax.persistence.GeneratedValue;
import javax.persistence.GenerationType;
import javax.persistence.Id;
import javax.persistence.JoinColumn;
import javax.persistence.ManyToOne;
import java.io.Serializable;

@Entity
@Table(name="tb_student")
public class Student implements Serializable {
```

```
    @Id
    @GeneratedValue(strategy = GenerationType.IDENTITY)
    private int id;
    private String name ;
    private String address ;
    private int age ;
    private char sex;
    // 学生与班级是多对一的关系，这里配置的是双向关联
    @ManyToOne(fetch=FetchType.LAZY,
        targetEntity=Clazz.class
        )
    @JoinColumn(name="clazzId", referencedColumnName="code")
    private Clazz clazz ;
    public Student() {

    }
    public Student(String name, String address, int age, char sex,
        Clazz clazz) {
        super();
        this.name = name;
        this.address = address;
        this.age = age;
        this.sex = sex;
        this.clazz = clazz;
    }
    // 省略 set/get 方法……
}
```

程序清单：codes/04/springdatajpaquerytest/src/main/java/org/fkit/springdatajpaquerytest/bean/Clazz.java

```
import java.util.HashSet;
import java.util.Set;
import javax.persistence.Entity;
import javax.persistence.FetchType;
import javax.persistence.GeneratedValue;
import javax.persistence.GenerationType;
import javax.persistence.Id;
import javax.persistence.OneToMany;
import java.io.Serializable;

@Entity
@Table(name="tb_clazz")
public class Clazz implements Serializable {
    @Id
    @GeneratedValue(strategy=GenerationType.IDENTITY)
    private int code ;
    private String name ;
    // 班级与学生是一对多的关联
    @OneToMany(
            fetch=FetchType.LAZY,
            targetEntity=Student.class,
            mappedBy="clazz"
            )
    private Set<Student> students = new HashSet<>();

    // 省略构造器和 set/get 方法……
}
```

从上述代码可以看出，班级与学生是一对多的关系，一个班级可以有多名学生，此处做的是双向关联，既在班级对象中关联了学生对象，在学生对象中也关联了班级对象。关于持久化类关联映射的具体配置，请参考疯狂软件系列之《轻量级 Java EE 企业应用实战》。

④ 定义数据访问层接口。

在 org.fkit.springdatajpaquerytest.repository 包下新建一个接口，命名为 StudentRepository，让该接口继承 JpaRepository 接口，以持久化对象 Student 作为 JpaRepository 的第一个类型参数，表示当前所操作的持久化对象类型，Integer 作为 JpaRepository 的第二个类型参数，用于指定 ID 类型，同时创建一个接口名称是 ClazzRepository，继承 JpaRepository 的接口，用于访问班级信息的数据。完整代码如下。

程序清单：codes/04/springdatajpaquerytest/src/main/java/org/fkit/springdatajpaquerytest/repository/ClazzRepository.java

```java
package org.fkit.springdatajpaquerytest.repository;

import org.fkit.springdatajpaquerytest.bean.Clazz;
import org.springframework.data.jpa.repository.JpaRepository;

public interface ClazzRepository extends JpaRepository<Clazz, Integer> {

}
```

以上数据访问层接口用于对班级表进行相关的 CRUD 操作。

程序清单：codes/04/springdatajpaquerytest/src/main/java/org/fkit/springdatajpaquerytest/repository/StudentRepository.java

```java
package org.fkit.springdatajpaquerytest.repository;
import java.util.List;
import java.util.Map;
import org.fkit.springdatajpaquerytest.bean.Student;
import org.springframework.data.jpa.repository.JpaRepository;
import org.springframework.data.jpa.repository.Modifying;
import org.springframework.data.jpa.repository.Query;
import org.springframework.data.repository.query.Param;

public interface StudentRepository extends JpaRepository<Student, Integer> {

    /**
     * 根据班级名称查询这个班级所有学生的信息
     * 相当于 JPQL 语句：select s from Student s where s.clazz.name = ?1
     * @param clazzName
     * @return
     */
    List<Student> findByClazz_name(String clazzName);

    /**
     * @Query 写法
     * 根据班级名称查询这个班级所有学生的信息
     * ?1 此处使用的是参数的位置，代表的是第一个参数
     * 此写法与 findByClazz_name 方法实现的功能完全一致
     * */
    @Query("select s from Student s where s.clazz.name = ?1")
    List<Student> findStudentsByClazzName(String clazzName);
    /**
     * 使用@Query 注解的形式，查询某个班级下所有学生的姓名和性别
     * @param clazzName
     * @return
     */
    @Query("select new Map(s.name as name, s.sex as sex)from Student s where s.clazz.name = ?1")
    List<Map<String, Object>> findNameAndSexByClazzName(String clazzName);

    /**
     * 使用@Query 注解的形式，查询某个班级下某种性别的所有学生的姓名
     * 上面方法用的是参数的位置来查询的，Spring Data JPA 中还支持用
```

```
     *   名称来匹配查询,使用格式":参数名称" 引用
     * @param clazzName
     * @return
     */
    @Query("select s.name from Student s  where s.clazz.name = :clazzName and s.sex = :sex )
    List<String> findNameByClazzNameAndSex(@Param("clazzName")String clazzName,
@Param("sex")char sex);
    /**
     *   使用@Query 注解的形式,查询某个学生属于哪个班级
     * @param stuName
     * @return
     */
    @Query("select c.name from Clazz c inner join c.students s where s.name = ?1 ")
    String findClazzNameByStuName(String stuName);
    /**
     *  执行更新查询,使用@Query 与@Modifying 可以执行更新操作
     *  例如删除牛魔王这个学生
     * */
    @Modifying
    @Query("delete from Student s where s.name = ?1")
    int deleteStuByStuName(String stuName);
}
```

➢ 从上述代码中可以看出,List<Student> findByClazz_name(String clazzName)方法是关联查询的方法,关联的属性可以用下画线 "_" 连接,此方法定义在学生对象的数据访问层接口下,相当于 JPQL 语句:select s from Student s where s.clazz.name =?1。

➢ @Query("select s from Student s where s.clazz.name = ?1")定义在方法 List<Student> findStudentsByClazzName(String clazzName)上,与 List<Student> findByClazz_name (String clazzName) 方法的功能是完全一致的,只是这里采用的是@Query 注解的形式进行查询。@Query 注解中可以直接定义 JPQL 语句进行数据的访问操作,@Query 查询摆脱了像命名查询那样的约束,将查询直接在相应的接口方法上声明,结构更为清晰,这是 Spring Data 的特有实现,非常方便。JPQL 语句中 "?1" 指代的是取方法形参列表中第 1 个参数值,1 代表的是参数位置,以此类推。

➢ @Query("select new Map(s.name as name, s.sex as sex) "
+ "from Student s where s.clazz.name = ?1")
List<Map<String,Object>> findNameAndSexByClazzName(String clazzName);
此方法使用 JPQL 语句直接返回 List<Map<String, Object>> 对象。

➢ @Query("select s.name from Student s where s.clazz.name = :clazzName and s.sex = :sex ")
List<String> findNameByClazzNameAndSex(@Param("clazzName")String clazzName,
    @Param("sex")char sex)
之前的方法是使用参数的位置来获取参数的值的,除此之外,Spring Data JPA 还支持用名称来匹配获取参数的值,使用格式为 ":参数名称",如上面方法所示,@Param 注解用于声明参数的名称,":参数名称" 用于提取对应参数的值。

➢ @Query("select c.name from Clazz c inner join c.students s where s.name = ?1 ")
String findClazzNameByStuName(String stuName)
此方法,实现了关联查询,查询某个学生所属的班级的名称。

➢ @Modifying
@Query("delete from Student s where s.name = ?1")

```
    int deleteStuByStuName (String stuName)
```
在此方法中，Spring Data JPA 支持使用@Modifying 和@Query 注解组合更新查询操作，上述方法即使用了此操作删除了某个学生数据。

⑤ 定义业务层类。

> 程序清单：codes/04/springdatajpaquerytest/src/main/java/org/fkit/
> springdatajpaquerytest/service/SchoolService.java

```java
import java.util.ArrayList;
import java.util.HashMap;
import java.util.List;
import java.util.Map;
import javax.annotation.Resource;
import org.fkit.springdatajpaquerytest.bean.Clazz;
import org.fkit.springdatajpaquerytest.bean.Student;
import org.fkit.springdatajpaquerytest.repository.ClazzRepository;
import org.fkit.springdatajpaquerytest.repository.StudentRepository;
import org.springframework.stereotype.Service;
import org.springframework.transaction.annotation.Transactional;
@Service
public class SchoolService {
    // 注入数据访问层接口对象
    @Resource
    private StudentRepository studentRepository;
    @Resource
    private ClazzRepository clazzRepository;
    @Transactional
    public void saveClazzAll(List<Clazz> clazzs) {
        clazzRepository.saveAll(clazzs);
    }
    @Transactional
    public void saveStudentAll(List<Student> students) {
        studentRepository.saveAll(students);
    }
    public List<Map<String, Object>> getStusByClazzName(String clazzName) {
        // 使用"_" 和 @Query 查询方式结果一致
        List<Student> students = studentRepository.findByClazz_name(clazzName);
        // List<Student> students = studentRepository.findStudentsByClazzName(clazzName);
        List<Map<String, Object>> results = new ArrayList<>();
        // 遍历查询出的学生对象，提取姓名、年龄、性别信息
        for(Student student:students){
            Map<String , Object> stu = new HashMap<>();
            stu.put("name", student.getName());
            stu.put("age", student.getAge());
            stu.put("sex", student.getSex());
            results.add(stu);
        }
        return results;
    }
    public List<Map<String, Object>> findNameAndSexByClazzName(String clazzName) {
        return studentRepository.findNameAndSexByClazzName(clazzName);
    }
    public List<String> findNameByClazzNameAndSex(
            String clazzName, char sex) {
        return studentRepository.findNameByClazzNameAndSex(clazzName, sex);
    }
    public String findClazzNameByStuName(String stuName) {
```

```java
        return studentRepository.findClazzNameByStuName(stuName);
    }
    @Transactional
    public int deleteStuByStuName(String stuName) {
        return studentRepository.deleteStuByStuName(stuName);
    }
}
```

在业务层中需要注入数据访问层对象，在上述代码中是通过@Resources 注解将 StudentRepositor 接口以及 ClazzRepository 接口对应的实现类对象注入进来的。

⑥ 定义控制器类。

在 org.fkit.springdatajpaquerytest.controller 包中新建一个控制器类，命名为 StudentController，完整代码如下。

程序清单：/springdatajpaquerytest/src/main/java/org/fkit/springdatajpaquerytest/controller/StudentController.java

```java
import java.util.ArrayList;
import java.util.List;
import java.util.Map;
import javax.annotation.Resource;
import org.fkit.springdatajpaquerytest.bean.Clazz;
import org.fkit.springdatajpaquerytest.bean.Student;
import org.fkit.springdatajpaquerytest.service.SchoolService;
import org.springframework.web.bind.annotation.RequestMapping;
import org.springframework.web.bind.annotation.RestController;

@RestController
@RequestMapping("/student")
public class StudentController {

// 注入 ShcoolService
    @Resource
    private SchoolService schoolService;
    @RequestMapping("/save")
    public String save() {

        Clazz clazz1 = new Clazz("疯狂java开发1班");
        Clazz clazz2 = new Clazz("疯狂java开发2班");
        // 保存班级对象数据
        List<Clazz> clazzs = new ArrayList<>();
        clazzs.add(clazz1);
        clazzs.add(clazz2);
        schoolService.saveClazzAll(clazzs);

        Student swk = new Student("孙悟空", "广州", 700, '男', clazz1);
        Student zzj = new Student("蜘蛛精", "广州", 700, '女', clazz1);
        Student nmw = new Student("牛魔王", "广州", 500, '男', clazz2);

        List<Student> students = new ArrayList<>();
        students.add(swk);
        students.add(zzj);
        students.add(nmw);
        schoolService.saveStudentAll(students);
        return "保存学生对象成功";
    }
    /**
     * 查询某个班级所有学生的姓名、年龄、性别
     * @param clazzName
     * @return
     */
    @RequestMapping("/getClazzStus")
```

```java
public List<Map<String, Object>> getClazzStus(String clazzName){
    return schoolService.getStusByClazzName(clazzName);
}
/**
 * 查询某个班级所有学生的姓名、性别
 * @param clazzName
 * @return
 */
@RequestMapping("/findNameAndSexByClazzName")
public List<Map<String, Object>> findNameAndSexByClazzName(String clazzName){
    return schoolService.findNameAndSexByClazzName(clazzName);
}
/**
 * 查询某个班级某种性别的所有学生的姓名
 * @param clazzName
 * @return
 */
@RequestMapping("/findNameByClazzNameAndSex")
public List<String> findNameByClazzNameAndSex(String clazzName , Character sex){
    return schoolService.findNameByClazzNameAndSex(clazzName , sex);
}
/**
 * 查询某个学生属于哪个班级
 * @param clazzName
 * @return
 */
@RequestMapping("/findClazzNameByStuName")
public String findClazzNameByStuName(String stuName){
    return schoolService.findClazzNameByStuName(stuName);
}
/**
 * 删除某个学生对象
 * @param clazzName
 * @return
 */
@RequestMapping("/deleteStuByStuName")
public String deleteStuByStuName(String stuName){
    return "删除数据："+schoolService.deleteStuByStuName(stuName);
}
```

⑦ 测试应用。

启动 MySQL 数据库，继续使用之前在数据库中创建的 springdatajpaquery 数据库，然后在 org.fkit.springdatajpaquerytest 包中新建 App.java 启动类，App.java 和之前的项目一致，此处不再赘述。右击该类运行 main 方法。Spring Boot 项目启动后，先去数据库查看是否成功自动创建了持久化类对应的 tb_student 和 tb_clazz 表。

测试添加学生和班级信息，在浏览器中输入如下地址。

```
http://127.0.0.1:8080/student/save
```

请求会提交给 StudentController 类的 save 方法进行处理，执行完成返回"保存学生对象成功"，查看数据库中的数据如图 4.19 和图 4.20 所示。

图 4.19 班级表的数据

图 4.20 学生表的数据

测试根据班级名称查询班级的学生信息，在浏览器中输入如下地址。

`http://127.0.0.1:8080/student/getClazzStus?clazzName=疯狂java开发1班`

请求会提交给 StudentController 类的 getClazzStus 方法进行处理，执行完成返回查询出的学生信息，如图 4.21 所示。

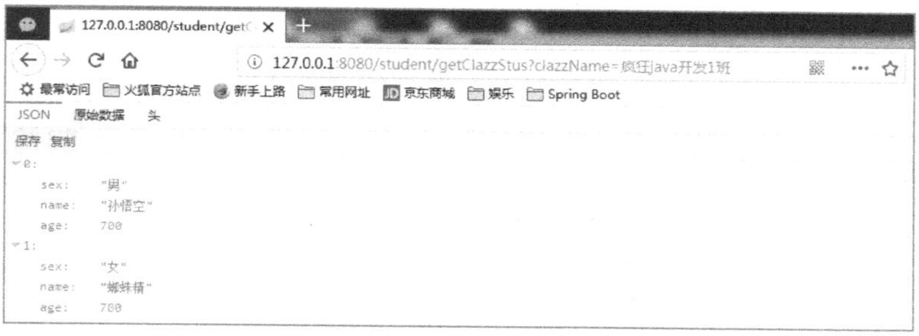

图 4.21　根据班级名称查询班级的学生信息

测试根据班级名称查询班级的学生姓名和性别，在浏览器中输入如下地址。

`http://127.0.0.1:8080/student/findNameAndSexByClazzName?clazzName=疯狂java开发1班`

请求会提交给 StudentController 类的 findNameAndSexByClazzName 方法进行处理，执行完成返回查询出的学生姓名和性别，如图 4.22 所示。

图 4.22　根据班级名称查询班级下的学生姓名和性别

测试查询某个班级的女学生，在浏览器中输入如下地址。

`http://127.0.0.1:8080/student/findNameByClazzNameAndSex?clazzName=疯狂java开发1班&sex=女`

请求会提交给 StudentController 类的 findNameAndSexByClazzName 方法进行处理，执行完成返回查询出的某个班级的女学生，如图 4.23 所示。

图 4.23　查询某个班级的女学生

测试查询某个学生属于哪个班级，在浏览器中输入如下地址。

`http://127.0.0.1:8080/student/findClazzNameByStuName?stuName=蜘蛛精`

请求会提交给 StudentController 类的 findClazzNameByStuName 方法进行处理，执行完成返回查询出的班级名称，如图 4.24 所示。

图 4.24　查询某个学生属于哪个班级

测试删除某个学生对象，在浏览器中输入如下地址。

http://127.0.0.1:8080/student/deleteStuByStuName?stuName=牛魔王

请求会提交给 StudentController 类的 deleteStuByStuName 方法进行处理，执行完成返回"删除数据：1"，查询数据库如图 4.25 所示。

图 4.25　删除某个学生后的数据

示例：@NamedQuery 查询

Spring Data JPA 支持使用 JPA 的 NameQuery 定义查询操作，即一个名称映射一个查询语句，接下来以示例的方式进行讲解。

① 创建一个新的 Maven 项目，命名为 springdatajpanamequerytest，修改 pom.xml 文件配置，全局配置文件的配置请参考"关联查询和@Query 查询"示例，此处不再赘述。

② 创建两个持久化类，Student.java 和 Clazz.java，其中 Clazz.java 代码请参考"关联查询和@Query 查询"示例，在 Student.java 中定义 NameQuery 查询，代码如下。

程序清单：codes/04/springdatajpanamequerytest/src/main/java/org/fkit/springdatajpanamequerytest/bean/Student.java

```java
import javax.persistence.Entity;
import javax.persistence.FetchType;
import javax.persistence.GeneratedValue;
import javax.persistence.GenerationType;
import javax.persistence.Id;
import javax.persistence.JoinColumn;
import javax.persistence.ManyToOne;
import javax.persistence.NamedQueries;
import javax.persistence.NamedQuery;
import javax.persistence.Table;
import java.io.Serializable;

@Entity
@Table(name="tb_student")
// 查询班级的学生信息
@NamedQuery(name="Student.findStudentsByClazzName", query="select s from Student s where s.clazz.name = ?1")
public class Student implements Serializable{
    @Id
    @GeneratedValue(strategy = GenerationType.IDENTITY)
    private int id;
    private String name ;
    private String address ;
    private int age ;
    private char sex;
    // 学生与班级是多对一的关系，这里配置的是双向关联
    @ManyToOne(fetch=FetchType.LAZY,
```

```
            targetEntity=Clazz.class
        )
    @JoinColumn(name="clazzId", referencedColumnName="code")
    private Clazz clazz ;
    public Student() {

    }
    public Student(String name, String address, int age, char sex,
            Clazz clazz) {
        super();
        this.name = name;
        this.address = address;
        this.age = age;
        this.sex = sex;
        this.clazz = clazz;
    }
    // 省略set/get方法……
}
```

其中@NamedQuery(name="Student.findStudentsByClazzName", query="select s from Student s where s.clazz.name = ?1")即定义了方法名称findStudentsByClazzName到query中的查询语句的关系，NameQuery语句应该放到要查询的实体上，并且名称是"实体的类名.方法名称"。

③ 定义数据访问层接口。

在org.fkit.springdatajpanamequerytest.repository包下创建两个数据访问层接口：ClazzRepository和StudentRepository，ClazzRepository接口参照"关联查询和@Query查询"示例的代码，StudentRepository接口的代码如下。

程序清单：codes/04/springdatajpanamequerytest/src/main/java/org/fkit/springdatajpanamequerytest/repository/StudentRepository.java

```java
import java.util.List;
import org.fkit.springdatajpanamequerytest.bean.Student;
import org.springframework.data.jpa.domain.Specifications;
import org.springframework.data.jpa.repository.JpaRepository;
import org.springframework.data.jpa.repository.JpaSpecificationExecutor;
public interface StudentRepository extends JpaRepository<Student, Integer> {
    /**
     * 查询班级的所有学生
     * @param clazzName
     * @return
     */
    List<Student> findStudentsByClazzName(String clazzName);
}
```

④ 定义业务层类。

程序清单：codes/04/springdatajpanamequerytest/src/main/java/org/fkit/springdatajpanamequerytest/service/ShcoolService.java

```java
import java.util.ArrayList;
import java.util.HashMap;
import java.util.List;
import java.util.Map;
import javax.annotation.Resource;
import org.fkit.springdatajpanamequerytest.bean.Clazz;
import org.fkit.springdatajpanamequerytest.bean.Student;
import org.fkit.springdatajpanamequerytest.repository.ClazzRepository;
import org.fkit.springdatajpanamequerytest.repository.StudentRepository;
import org.springframework.stereotype.Service;
import org.springframework.transaction.annotation.Transactional
```

```java
@Service
public class ShcoolService {
    // 注入数据访问层接口对象
    @Resource
    private StudentRepository studentRepository;
    @Resource
    private ClazzRepository clazzRepository;

    @Transactional
    public void saveClazzAll(List<Clazz> clazzs) {
        clazzRepository.saveAll(clazzs);
    }
    @Transactional
    public void saveStudentAll(List<Student> students) {
        studentRepository.saveAll(students);
    }

    public List<Map<String, Object>> getStusByClazzName(String clazzName) {
        // 查询班级的所有学生
        List<Student> students = studentRepository.findStudentsClazzName(clazzName);
        List<Map<String, Object>> results = new ArrayList<>();
        // 遍历查询出的学生对象，提取姓名、年龄、性别信息
        for(Student student:students){
            Map<String , Object> stu = new HashMap<>();
            stu.put("name", student.getName());
            stu.put("age", student.getAge());
            stu.put("sex", student.getSex());
            results.add(stu);
        }
        return results;
    }
}
```

⑤ 定义控制器类。

在 org.fkit.springdatajpanamequerytest.controller 包中新建一个控制器类，命名为 StudentController，其代码如下。

程序清单：codes/04/springdatajpanamequerytest/src/main/java/org/fkit/
springdatajpanamequerytest/controller/StudentController.java

```java
import java.util.ArrayList;
import java.util.List;
import java.util.Map;
import javax.annotation.Resource;
import org.fkit.springdatajpanamequerytest.bean.Clazz;
import org.fkit.springdatajpanamequerytest.bean.Student;
import org.fkit.springdatajpanamequerytest.service.ShcoolService;
import org.springframework.web.bind.annotation.RequestMapping;
import org.springframework.web.bind.annotation.RestController;

@RestController
@RequestMapping("/student")
public class StudentController {
    @Resource
    private ShcoolService shcoolService;
    @RequestMapping("/save")
    public String save() {

        Clazz clazz1 = new Clazz("疯狂java开发1班");
        Clazz clazz2 = new Clazz("疯狂java开发2班");
        // 保存班级对象数据
        List<Clazz> clazzs = new ArrayList<>();
        clazzs.add(clazz1);
```

```
            clazzs.add(clazz2);
            shcoolService.saveClazzAll(clazzs);

            Student swk = new Student("孙悟空", "广州", 700, '男', clazz1);
            Student zzj = new Student("蜘蛛精", "广州", 700, '女', clazz1);
            Student nmw = new Student("牛魔王", "广州", 500, '男', clazz2);

            List<Student> students = new ArrayList<>();
            students.add(swk);
            students.add(zzj);
            students.add(nmw);
            shcoolService.saveStudentAll(students);
            return "保存学生对象成功";
        }
        /**
         * 查询某个班级所有学生的姓名、年龄、性别
         * @param clazzName
         * @return
         */
        @RequestMapping("/getClazzStus")
        public List<Map<String, Object>> getClazzStus(String clazzName){
            return shcoolService.getStusByClazzName(clazzName);
        }
    }
```

⑥ 测试应用。

启动 MySQL 数据库，如果继续使用之前在数据库中创建的 springdatajpaquery 数据库，应先删除之前的数据库表 tb_student 和 tb_clazz，以免影响本示例的测试，然后在 org.fkit.springdatajpanamequerytest 包下新建 App.java 启动类。App.java 和之前的项目一致，此处不再赘述。右击该类运行 main 方法。Spring Boot 项目启动后，JPA 会在数据库中自动创建持久化类对应的 tb_student 和 tb_clazz 表。

测试添加学生和班级信息，在浏览器中输入如下地址。

```
http://127.0.0.1:8080/student/save
```

请求会提交给 StudentController 类的 save 方法进行处理，执行完成返回"保存学生对象成功"，详细数据与"关联查询和@Query 查询"示例一致，接下来测试查询某个班级下的所有学生信息，在浏览器中输入：

```
http://127.0.0.1:8080/student/getClazzStus?clazzName=疯狂java开发1班
```

请求会提交到 StudentController 类的 getClazzStus 方法进行处理，执行完成返回查询出的学生对象，如图 4.26 所示。

图 4.26　查询班级的学生信息

示例：Specification 查询

JPA 允许基于 Criteria 对象进行按条件查询，而 Spring Data JPA 提供了一个 Specification 接口，Specification 接口封装了 JPA 的 Criteria 查询条件，从而可以通过此接口更加方便地使用 Criteria 查询，Specification 接口的源代码如下。

程序清单 org/springframework/data/jpa/domain/Specification.java

```java
package org.springframework.data.jpa.domain;
import javax.persistence.criteria.CriteriaBuilder;
import javax.persistence.criteria.CriteriaQuery;
import javax.persistence.criteria.Predicate;
import javax.persistence.criteria.Root;
public interface Specification<T> {
    Predicate toPredicate(Root<T> root, CriteriaQuery<?> query, CriteriaBuilder cb);
}
```

Specification 接口提供了一个 toPredicate 方法用来构造查询条件。值得注意的是：如果自己定义的数据访问层接口希望使用 Specification 接口的规范，则必须实现 JpaSpecificationExecutor 接口，JpaSpecificationExecutor 接口不属于 Repository 体系，它实现了一组 JPA Criteria 查询相关的方法，查看 JpaSpecificationExecutor 接口的源代码代码如下。

程序清单 org/springframework/data/jpa/repository/JpaSpecificationExecutor.java

```java
package org.springframework.data.jpa.repository;
import java.util.List;
import org.springframework.data.domain.Page;
import org.springframework.data.domain.Pageable;
import org.springframework.data.domain.Sort;
import org.springframework.data.jpa.domain.Specification;
public interface JpaSpecificationExecutor<T> {
    // 根据 Specification 接口封装的查询条件，查询出单一实例
    T findOne(Specification<T> spec);
    // 根据 Specification 接口封装的查询条件，查询出满足条件的所有对象数据，返回的是 List 集合
    List<T> findAll(Specification<T> spec);
    /** 根据 Specification 接口封装的查询条件以及 Pageable 封装的分页及排序规则查询出满足条
     件的对象数据，返回的是 Page 对象，Page 对象中封装了查询出的数据信息以及分页信息
     */
    Page<T> findAll(Specification<T> spec, Pageable pageable);
    /** 根据 Specification 接口封装的查询条件以及 Sort 封装的排序规则查询出满足条件的所有对
     象数据，返回的是 List 集合
     */
    List<T> findAll(Specification<T> spec, Sort sort);
    // 根据 Specification 接口封装的查询条件，查询出满足此条件的数据总数
    long count(Specification<T> spec);
}
```

从 JpaSpecificationExecutor 的方法中可以看出，一般通过创建 Specification 的匿名内部类对象来封装 Criteria 条件查询信息，然后交由 JpaSpecificationExecutor 提供的方法进行相关数据的查询操作。

➢ T findOne(Specification<T> spec)：根据 Specification 接口封装的查询条件，查询单一实例。

➢ List<T> findAll(Specification<T> spec)：根据 Specification 接口封装的查询条件，查询满足条件的所有对象数据，返回的是 List 集合。

➢ Page<T> findAll(Specification<T> spec, Pageable pageable)：根据 Specification 接口封装的查询条件以及 Pageable 封装的分页及排序规则查询出满足条件的对象数据，返回

的是 Page 对象，Page 对象中封装了查询出的数据信息以及分页信息。
- List<T> findAll(Specification<T> spec, Sort sort)：根据 Specification 接口封装的查询条件以及 Sort 封装的排序规则查询出满足条件的所有对象数据，返回的是 List 集合。
- long count(Specification<T> spec)：根据 Specification 接口封装的查询条件，查询出满足此条件的数据总数。

所以在实际项目开发中，如果希望使用 Specification 查询，数据访问层的定义通常如以下代码所示。

```
public interface StudentRepository extends JpaRepository<Student, Integer>,
JpaSpecificationExecutor<Student>
{

}
```

接下来以示例的方式来详细讲解使用 Specification 的查询、分页、动态查询等操作。

① 创建一个新的 Maven 项目，命名为 springdatajpaspecificationtest，修改 pom.xml 文件请参考"@NamedQuery 查询"示例配置，内容完全一样，此处不再赘述。

② 配置基本属性。

在 src/main/resources 包下新建一个全局配置文件，命名为 application.properties，同时在 MySQL 数据库中执行如下命令。

```
CREATE DATABASE springdatajpaspecification; #创建一个新的数据库，名称是
springdatajpaspecification
```

在 application.properties 配置文件中配置数据源和 jpa 相关的属性，详细代码如下。

```
############################################################
### 数据源信息配置
############################################################
# 数据库地址
spring.datasource.url = jdbc:mysql://localhost:3306/springdatajpaspecification
# 用户名
spring.datasource.username = root
# 密码
spring.datasource.password =
# 数据库驱动
spring.datasource.driverClassName = com.mysql.jdbc.Driver
# 指定连接池中最大的活跃连接数
spring.datasource.max-active=20
# 指定连接池最大的空闲连接数
spring.datasource.max-idle=8
# 指定必须保持连接的最小值
spring.datasource.min-idle=8
# 指定启动连接池时，初始建立的连接数量
spring.datasource.initial-size=10


############################################################
### JPA 持久化配置
############################################################
# 指定数据库的类型
spring.jpa.database = MySQL
# 指定是否需要在日志中显示 SQL 语句
spring.jpa.show-sql = true
# 指定自动创建|更新|验证数据库表结构等配置，配置成 update
# 表示如果数据库中存在持久化类对应的表就不创建，不存在就创建对应的表
spring.jpa.hibernate.ddl-auto = update
```

```
# 指定命名策略
spring.jpa.hibernate.naming-strategy = org.hibernate.cfg.ImprovedNamingStrategy
# 指定数据库方言
spring.jpa.properties.hibernate.dialect = org.hibernate.dialect.MySQL5Dialect
```

③ 创建持久化类。

在org.fkit.springdatajpaspecificationtest.bean包下新建四个包,分别是bean(放置持久化类)、controller(控制器)、repository(定义数据访问接口的包)、service(业务逻辑处理类),在bean包中创建两个持久化类:Student.java 和 Clazz.java,代码与"关联查询和@Query 查询"示例代码一致,读者可参考"关联查询和@Query 查询"示例中的代码。

④ 定义数据访问层接口。

在org.fkit.springdatajpaspecificationtest.repository包下新建一个接口,命名为StudentRepository,让该接口继承JpaRepository接口,以持久化对象Student作为JpaRepository的第一个类型参数,表示当前所操作的持久化对象类型,Integer作为JpaRepository的第二个类型参数,用于指定ID 类型,同时创建一个接口,命名为ClazzRepository,继承 JpaRepository 的接口,用于访问班级信息的数据。完整代码如下。

程序清单:codes/04/springdatajpaspecificationtest/src/main/java/org/fkit/
springdatajpaspecificationtest/repository/ClazzRepository.java

```java
import org.fkit.springdatajpaspecificationtest.bean.Clazz;
import org.springframework.data.jpa.repository.JpaRepository;
import org.springframework.data.jpa.repository.JpaSpecificationExecutor;

public interface ClazzRepository extends JpaRepository<Clazz , Integer> , JpaSpecificationExecutor<Clazz>{

}
```

以上数据访问层接口用于对班级表进行相关的 CRUD 操作,同时由于实现了JpaSpecificationExecutor 接口,ClazzRepository 接口也将拥有 JpaSpecificationExecutor 接口提供的功能。

程序清单:codes/04/springdatajpaspecificationtest/src/main/java/org/fkit/
springdatajpaspecificationtest/repository/StudentRepository.java

```java
import org.fkit.springdatajpaspecificationtest.bean.Student;
import org.springframework.data.jpa.repository.JpaRepository;
import org.springframework.data.jpa.repository.JpaSpecificationExecutor;

public interface StudentRepository extends JpaRepository<Student , Integer> , JpaSpecificationExecutor<Student>
{

}
```

⑤ 定义业务层类。

程序清单:codes/04/springdatajpaspecificationtest/src/main/java/org/fkit/
springdatajpaspecificationtest/service/ShcoolService.java

```java
import java.util.ArrayList;
import java.util.HashMap;
import java.util.List;
import java.util.Map;
import javax.annotation.Resource;
import javax.persistence.criteria.CriteriaBuilder;
import javax.persistence.criteria.CriteriaQuery;
import javax.persistence.criteria.JoinType;
import javax.persistence.criteria.Path;
```

```java
import javax.persistence.criteria.Predicate;
import javax.persistence.criteria.Root;
import org.fkit.springdatajpaspecificationtest.bean.Clazz;
import org.fkit.springdatajpaspecificationtest.bean.Student;
import org.fkit.springdatajpaspecificationtest.repository.ClazzRepository;
import org.fkit.springdatajpaspecificationtest.repository.StudentRepository;
import org.springframework.data.domain.Page;
import org.springframework.data.domain.PageRequest;
import org.springframework.data.domain.Sort;
import org.springframework.data.jpa.domain.Specification;
import org.springframework.stereotype.Service;
import org.springframework.transaction.annotation.Transactional;
import org.springframework.util.StringUtils;
@Service
public class ShcoolService {
    // 注入数据访问层接口对象
    @Resource
    private StudentRepository studentRepository;
    @Resource
    private ClazzRepository clazzRepository;
    @Transactional
    public void saveClazzAll(List<Clazz> clazzs) {
        clazzRepository.saveAll(clazzs);
    }
    @Transactional
    public void saveStudentAll(List<Student> students) {
        studentRepository.saveAll(students);
    }

    /**
     * 根据性别查询学生信息
     * @param clazzName
     * @return
     */
    @SuppressWarnings("serial")
    public List<Map<String, Object>> getStusBySex(char sex) {
        List<Student> students = studentRepository.findAll(new Specification<Student>() {
            @Override
            public Predicate toPredicate(Root<Student> root, CriteriaQuery<?> query,
                    CriteriaBuilder cb) {
                // root.get("sex")表示获取sex这个字段名称,equal表示执行equal查询
                // 相当于 select s from Student s where s.sex = ?1
                Predicate p1 = cb.equal(root.get("sex"), sex);
                return p1;
            }
        });
        List<Map<String, Object>> results = new ArrayList<>();
        // 遍历查询出的学生对象,提取姓名、年龄、性别信息
        for(Student student:students){
            Map<String , Object> stu = new HashMap<>();
            stu.put("name", student.getName());
            stu.put("age", student.getAge());
            stu.put("sex", student.getSex());
            results.add(stu);
        }
        return results;
    }

    /**
     * 动态查询学生信息：可以根据学生对象的姓名（模糊匹配）、地址查询（模糊匹配）、性别、班级
     * 查询学生信息
     * 如果没有传输参数,默认查询所有的学生信息
```

```java
     * @param clazzName
     * @return
     */
    @SuppressWarnings("serial")
    public List<Map<String, Object>> getStusByDynamic(Student student) {
        List<Student> students = studentRepository.findAll(new Specification<Student>() {
            @Override
            public Predicate toPredicate(Root<Student> root, CriteriaQuery<?> query,
                    CriteriaBuilder cb) {
                // 本集合用于封装查询条件
                List<Predicate> predicates = new ArrayList<Predicate>();
                if(student!=null){
                    /** 是否传入用于查询的姓名   */
                    if(!StringUtils.isEmpty(student.getName())){
                        predicates.add(cb.like(root.<String> get("name"), "%" +
student.getName() + "%"));
                    }
                    /** 是否传入用于查询的地址   */
                    if(!StringUtils.isEmpty(student.getAddress())){
                        predicates.add(cb.like(root.<String> get("address"), "%" +
student.getAddress() + "%"));
                    }
                    /** 是否传入用于查询的性别   */
                    if(student.getSex() != '\0'){
                        predicates.add(cb.equal(root.<String> get("sex"), student.getSex()));
                    }
                    /** 是否传入用于查询的班级信息   */
                    if(student.getClazz()!=null && !StringUtils.isEmpty
(student.getClazz().getName())){
                        root.join("clazz", JoinType.INNER);
                        Path<String> clazzName = root.get("clazz").get("name");
                        predicates.add(cb.equal(clazzName, student.getClazz().getName()));
                    }
                }
                return  query.where(predicates.toArray(new  Predicate[predicates.
size()])).getRestriction();
            }
        });
        List<Map<String, Object>> results = new ArrayList<>();
        // 遍历查询出的学生对象，提取姓名、年龄、性别信息
        for(Student stu :students){
            Map<String , Object> stuMap = new HashMap<>();
            stuMap.put("name", stu.getName());
            stuMap.put("age", stu.getAge());
            stuMap.put("sex", stu.getSex());
            stuMap.put("address", stu.getAddress());
            stuMap.put("clazzName", stu.getClazz().getName());
            results.add(stuMap);
        }
        return results;
    }
    /**
     * 分页查询某个班级的学生信息
     * @param clazzName 代表班级名称
     * @param pageIndex 代表当前查询第几页
     * @param pageSize 代表每页查询的最大数据量
     * @return
     */
    @SuppressWarnings("serial")
    public Page<Student> getStusByPage(String clazzName , int pageIndex , int
pageSize ) {
```

```java
        // 指定排序参数对象：根据 id, 进行降序查询
        Sort sort = new Sort(Sort.Direction.DESC, "id");
        // 分页查询学生信息，返回分页实体对象数据
        // pages 对象中包含了查询出来的数据信息以及与分页相关的信息
        Page<Student> pages = studentRepository.findAll(new Specification<Student>() {
            @Override
            public Predicate toPredicate(Root<Student> root, CriteriaQuery<?> query,
                    CriteriaBuilder cb) {
                root.join("clazz", JoinType.INNER);
                Path<String> cn = root.get("clazz").get("name");
                Predicate p1 = cb.equal(cn, clazzName);
                return p1 ;
            }
        }, PageRequest.of(pageIndex-1, pageSize, sort));
        return pages;
    }
}
```

在业务层中需要注入数据访问层对象，在上述代码中我们是通过@Resources 注解将 StudentRepository 接口以及 ClazzRepository 接口对应的实现类对象注入的，同时在业务层方法中定义了三个方法，分别实现了对班级信息的条件查询、动态 SQL 语句查询以及分页查询。

⑥ 定义分页的页面数据对象。

在 org.fkit.springdatajpaspecificationtest 包下新建一个子包，命名为 vo，在 vo 下新建一个 Java 类，命名为 PageData.java，此类用于封装分页查询出的数据信息，主要包含了当前页码（pageIndex）、满足查询条件下用于分页的数据总量（totalCount）、当前条件下总共可以分的总页数（pageSize）、当前页码展示的数据量（pageNum）以及查询出的数据信息（stuDatas）。详细代码如下。

程序清单：codes/04/springdatajpaspecificationtest/src/main/java/org/fkit/springdatajpaspecificationtest/vo/PageData.java

```java
import java.util.ArrayList;
import java.util.List;
import java.util.Map;
/**
 * 定义一个对象用于封装一页数据
 */
public class PageData {
    // 定义一个变量用于存放当前页码
    private int pageIndex;
    // 定义一个变量用于保存满足查询条件下用于分页的数据总量
    private long totalCount;
    // 定义一个变量用于保存当前条件下可以分的总页数
    private int pageSize ;
    // 定义一个变量用于保存当前页码查询出的数据总量
    private int pageNum;
    // 定义一个变量用于保存当前查询出来的学生信息
    private List<Map<String, Object>> stuDatas = new ArrayList<>();

    // 省略构造器和set/get 方法……
}
```

⑦ 定义控制器类。

在 org.fkit.springdatajpaspecificationtest.controller 包中新建一个控制器类，命名为 StudentController，其代码如下。

程序清单:codes/04/springdatajpaspecificationtest/src/main/java/org/fkit/
springdatajpaspecificationtest/controller/StudentController.java

```java
import java.util.ArrayList;
import java.util.HashMap;
import java.util.List;
import java.util.Map;
import javax.annotation.Resource;
import org.fkit.springdatajpaspecificationtest.bean.Clazz;
import org.fkit.springdatajpaspecificationtest.bean.Student;
import org.fkit.springdatajpaspecificationtest.service.ShcoolService;
import org.fkit.springdatajpaspecificationtest.vo.PageData;
import org.springframework.data.domain.Page;
import org.springframework.web.bind.annotation.RequestMapping;
import org.springframework.web.bind.annotation.RestController;

@RestController
@RequestMapping("/student")
public class StudentController {

    // 注入SchoolService
    @Resource
    private SchoolService schoolService;
    @RequestMapping("/save")
    public String save() {

        Clazz clazz1 = new Clazz("疯狂java开发1班");
        Clazz clazz2 = new Clazz("疯狂java开发2班");
        // 保存班级对象数据
        List<Clazz> clazzs = new ArrayList<>();
        clazzs.add(clazz1);
        clazzs.add(clazz2);
        shcoolService.saveClazzAll(clazzs);

        Student swk = new Student("孙悟空", "花果山", 700, '男', clazz1);
        Student zx = new Student("紫霞仙子", "盘丝洞", 500, '女', clazz1);
        Student zzb = new Student("至尊宝", "广州", 500, '男', clazz1);
        Student tsgz = new Student("铁扇公主", "火焰山", 500, '女', clazz2);
        Student nmw = new Student("牛魔王", "广州", 500, '男', clazz2);
        Student zzj = new Student("蜘蛛精", "广州", 700, '女', clazz2);

        List<Student> students = new ArrayList<>();
        students.add(swk);
        students.add(zx);
        students.add(zzb);
        students.add(tsgz);
        students.add(nmw);
        students.add(zzj);
        shcoolService.saveStudentAll(students);
        return "保存学生对象成功";
    }

    @RequestMapping("/getStusBySex")
    public List<Map<String, Object>> getStusBySex(char sex){
        return shcoolService.getStusBySex(sex);
    }
    // 动态查询学生信息
    // 可以根据学生对象的姓名(模糊匹配)、地址查询(模糊匹配)、性别、班级查询学生信息
    @RequestMapping("/getStusByDynamic")
    List<Map<String, Object>> getStusByDynamic(Student student) {
        return shcoolService.getStusByDynamic(student);
    }
    // 分页查询某个班级的学生信息
```

```
    @RequestMapping("/getStusByPage")
    PageData getStusByPage(String clazzName, int pageIndex, int pageSize ) {
        // 分页查询某个班级的学生信息
        Page<Student> page = shcoolService.getStusByPage(clazzName, pageIndex, pageSize);
        // 对查询出来的结果数据进行分析
        List<Student> students = page.getContent();
        List<Map<String, Object>> stuDatas = new ArrayList<>();
        for(Student stu :students){
            Map<String , Object> stuMap = new HashMap<>();
            stuMap.put("name",  stu.getName());
         stuMap.put("id", stu.getId());
            stuMap.put("age",  stu.getAge());
            stuMap.put("sex",  stu.getSex());
            stuMap.put("address", stu.getAddress());
            stuMap.put("clazzName",  clazzName);
            stuDatas.add(stuMap);
        }
        // 将分页查询出的结果数据进行分析
        // 然后把数据存入 PageData 对象中响应给浏览器展示
        PageData data = new PageData();
        data.setStuDatas(stuDatas);
        data.setPageIndex(page.getNumber()+1);
        data.setPageSize(page.getTotalPages());
        data.setTotalCount(page.getTotalElements());
        data.setPageNum(page.getSize());
        return data ;
    }
}
```

在控制器类中定义了四个方法。其中 save 方法用于保存数据，也作为初始化的测试数据使用。getStusBySex(char sex)方法用于根据性别查询学生信息。getStusByDynamic(Student student)方法用于根据 student 对象中的姓名（模糊匹配）、地址（模糊匹配）、性别、班级信息动态查询满足条件的学生信息，如果没有任何查询条件则查询出系统所有的学生信息。getStusByPage(String clazzName, int pageIndex, int pageSize)方法用于分页查询某个班级的学生信息，返回的是一个分页信息 Page<Student>，其中 clazzName 是指班级参数，pageIndex 是指查询的当前页码，pageSize 是指每页最多显示多少条数据。

⑧ 测试应用。

启动 MySQL 数据库，创建一个新的数据库，命名为 springdatajpaspecification，然后在 org.fkit.springdatajpaspecificationtest 包下新建 App.java 启动类，App.java 和之前的项目一致，此处不再赘述。右击该类运行 main 方法。Spring Boot 项目启动后，JPA 会在数据库中自动创建持久化类对应的 tb_student 和 tb_clazz 表。

测试添加学生和班级信息，在浏览器中输入如下地址：

http://127.0.0.1:8080/student/save

请求会提交到 StudentController 类的 save 方法进行处理，执行完成返回"保存学生对象成功"，查看数据库中的数据，如图 4.27 和图 4.28 所示。

图 4.27 班级表的数据

图 4.28 学生表的数据

测试根据性别查询学生信息，在浏览器中输入如下地址。

http://127.0.0.1:8080/student/getStusBySex?sex=女

请求会提交给 StudentController 类的 getStusBySex 方法进行处理，执行完成以后，查询出的学生信息如图 4.29 所示。

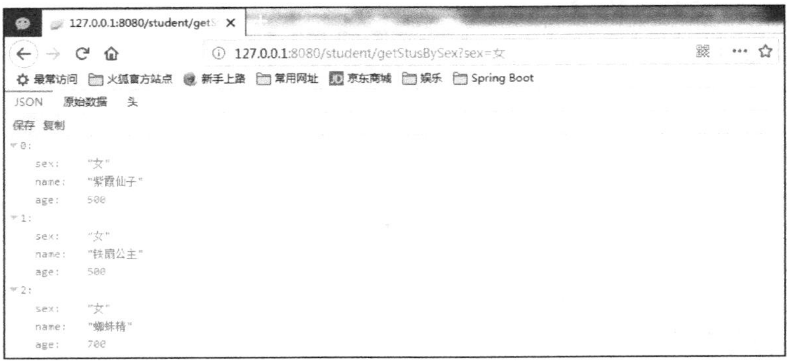

图 4.29　查询某个性别的学生信息

测试动态查询学生信息，可以根据学生对象的姓名（模糊匹配）、地址（模糊匹配）、性别、班级查询学生信息，在浏览器中输入如下地址。

http://127.0.0.1:8080/student/getStusByDynamic?clazz.name=疯狂java开发1班&sex=女

请求会提交给 StudentController 类的 getStusByDynamic 方法进行处理，此地址是通过学生的班级和性别来查询对应的学生信息的，执行完成以后，查询出的学生信息如图 4.30 所示。

图 4.30　根据班级名和性别动态查询学生信息

在浏览器中输入如下地址：

http://127.0.0.1:8080/student/getStusByDynamic?address=广州&sex=男

请求会提交给 StudentController 类的 getStusByDynamic 方法进行处理，这是通过学生的地址（模糊匹配）和性别来查询对应的学生信息的，执行完成以后，查询出的学生信息如图 4.31 所示。

测试分页查询某个班级下的学生信息，在浏览器中输入如下地址。

http://127.0.0.1:8080/student/getStusByPage?clazzName=疯狂java开发1班&pageIndex=1&pageSize=2

请求会提交给 StudentController 类的 getStusByPage 方法进行处理，地址是降序查询"疯狂 java 开发 1 班"的第 1 页数据，每页最多展示 2 条数据，执行完成以后，查询出的学生信息如图 4.32 所示。

图 4.31　根据地址和性别动态查询学生信息

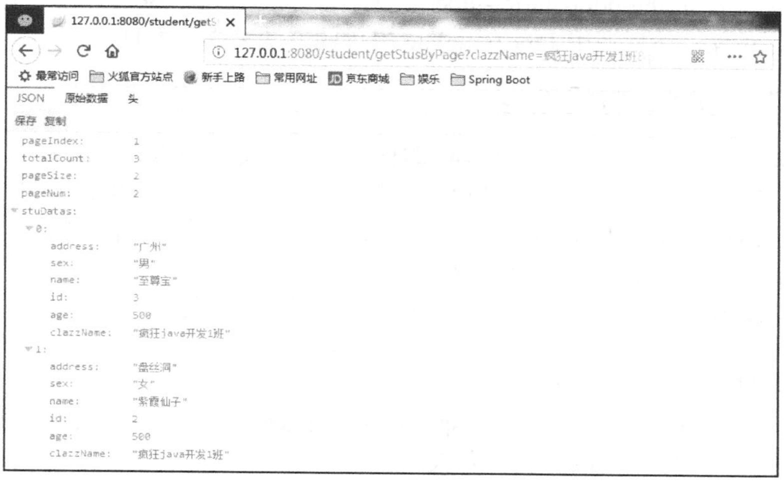

图 4.32　分页查询某个班级的第 1 页数据

从图 4.32 可以看到，pageIndex 表示这是第 1 页数据，totalCount 表示"疯狂 java 开发 1 班"当前总共有 3 条学生记录，pageSize 表示一页最多展示 2 条数据，pageNum 表示当前总共有 2 页数据，stuDatas 即查询出的当前页数据。

在浏览器中输入如下地址来查询第 2 页数据。

```
http://127.0.0.1:8080/student/getStusByPage?clazzName=疯狂 java 开发 1 班&pageIndex=2&pageSize=2
```

执行完成后的结果如图 4.33 所示。

图 4.33　分页查询某个班级的第 2 页数据

从图 4.33 中可以看到，pageIndex 表示这是第 2 页数据，totalCount 表示"疯狂 java 开发 1 班"当前总共有 3 条学生记录，pageSize 表示一页最多展示 2 条数据，pageNum 表示当前总共有 2 页数据，stuDatas 即是查询出的当前页数据。读者可自行更换条件进行测试。

4.3 Spring Boot 使用 JdbcTemplate

Spring 对数据库的操作在 JDBC 上做了深层次的封装，建立了一个 JDBC 存取框架 JdbcTemplate（JDBC 模板）。JdbcTemplate 设计的目的是为不同类型的 JDBC 操作提供模板方法，每个模板方法都能控制整个数据访问的过程，通过这种方式，可以在尽可能保持灵活性的情况下，将数据库存取的工作量降到最低，通过使用 Spring Boot 自动配置功能，在 Maven 配置文件中，需要增加 spring-boot-starter-jdbc 模块，即可在 Spring Boot 项目中通过注解轻松注入 JdbcTemplate 对象，然后调用 JdbcTemplate 提供的方法来操作数据库。

示例：JdbcTemplate 访问数据

JdbcTemplate 对象是 Spring 封装 JDBC 的核心对象，提供的常用方法如下。
- execute 方法：可以用于执行任何 SQL 语句，一般用于执行 DDL 语句。
- update 方法和 batchUpdate 方法：update 方法用于执行新增、修改、删除等语句；batchUpdate 方法用于执行批处理相关语句。
- query 方法和 queryFor*XXX* 方法：用于执行查询相关语句。
- call 方法：用于执行存储过程、函数相关语句。

接下来以项目示例的方法来详细介绍 JdbcTemplate。

创建一个新的 Maven 项目，命名为 springbootjdbctemplate。按照 Maven 项目的规范，在 src/main/下新建一个名为 resources 的文件夹。

① 修改 pom.xml 文件。

在 pom.xml 文件中增加 Web 开发依赖配置。

```xml
<dependency>
    <groupId>org.springframework.boot</groupId>
    <artifactId>spring-boot-starter-web</artifactId>
</dependency>
```

然后添加数据库驱动依赖配置。

```xml
<dependency>
    <groupId>mysql</groupId>
    <artifactId>mysql-connector-java</artifactId>
</dependency>
```

还需要添加 spring-boot-starter-jdbc，它是 Spring Boot 项目使用 JdbcTemplate 访问数据库的核心依赖配置。

```xml
<dependency>
    <groupId>org.springframework.boot</groupId>
    <artifactId>spring-boot-starter-jdbc</artifactId>
</dependency>
```

修改后的完整 pom.xml 文件如下。

程序清单： codes/04//springbootjdbctemplate/pom.xml

```
<project xmlns="http://maven.apache.org/POM/4.0.0" xmlns:xsi="http://www.w3.org/2001/XMLSchema-instance"
    xsi:schemaLocation="http://maven.apache.org/POM/4.0.0
```

```xml
http://maven.apache.org/xsd/maven-4.0.0.xsd">
    <modelVersion>4.0.0</modelVersion>
    <groupId>org.fkit</groupId>
    <artifactId>springbootjdbctemplate</artifactId>
    <version>0.0.1-SNAPSHOT</version>
    <packaging>jar</packaging>
    <name>springbootjdbctemplate</name>
    <url>http://maven.apache.org</url>
    <parent>
        <groupId>org.springframework.boot</groupId>
        <artifactId>spring-boot-starter-parent</artifactId>
        <version>2.0.0.RELEASE</version>
    </parent>
    <properties>
        <project.build.sourceEncoding>UTF-8</project.build.sourceEncoding>
        <project.reporting.outputEncoding>UTF-8</project.reporting.outputEncoding>
        <java.version>1.8</java.version>
    </properties>
    <dependencies>
        <!-- 添加 spring-boot-starter-web 依赖... -->
        <dependency>
            <groupId>org.springframework.boot</groupId>
            <artifactId>spring-boot-starter-web</artifactId>
        </dependency>
        <!-- 添加 spring-boot-starter-thymeleaf 模块依赖 -->
        <dependency>
            <groupId>org.springframework.boot</groupId>
            <artifactId>spring-boot-starter-thymeleaf</artifactId>
        </dependency>

        <!-- 添加 MySQL 依赖 -->
        <dependency>
            <groupId>mysql</groupId>
            <artifactId>mysql-connector-java</artifactId>
        </dependency>

        <!-- 添加 JDBC 依赖 -->
        <dependency>
            <groupId>org.springframework.boot</groupId>
            <artifactId>spring-boot-starter-jdbc</artifactId>
        </dependency>
    </dependencies>
</project>
```

❷ 配置基本属性。

在 src/main/resources 包下新建一个全局配置文件，命名为 application.properties，在该配置文件中配置数据源。

程序清单：codes/04//springbootjdbctemplate/src/main/resources/application.properties

```
######################################################
### 数据源信息配置
######################################################
# 连接地址
spring.datasource.url=jdbc:mysql://localhost:3306/jdbctemplate
# 用户名配置
spring.datasource.username=root
# 密码
spring.datasource.password=
# 数据库驱动
spring.datasource.driver-class-name=com.mysql.jdbc.Driver
```

③ 创建持久化类。

在 org.fkit.springbootjdbctemplate 包下新建 4 个包,分别是 bean(放置持久化类)、controller(控制器)、repository(定义数据访问接口的包)、service(业务逻辑处理类),在 bean 包中创建一个持久化类 User.java,其代码如下。

程序清单:codes/04/springbootjdbctemplate/src/main/java/org/fkit/springbootjdbctemplate/bean/User.java

```java
import java.io.Serializable;
public class User implements Serializable{
    private int id ;
    private String loginName ;
    private String username ;
    private String password;

    // 省略构造器和 set/get 方法……
}
```

④ 定义数据访问层接口。

在 org.fkit.springbootjdbctemplate.repository 包下新建一个接口,命名为 UserRepository.java,完整代码结构如下。

程序清单:codes/04/springbootjdbctemplate/src/main/java/org/fkit/springbootjdbctemplate/repository/UserRepository.java

```java
import java.sql.Connection;
import java.sql.PreparedStatement;
import java.sql.SQLException;
import java.sql.Statement;
import java.util.List;
import javax.annotation.Resource;
import org.fkit.springbootjdbctemplate.bean.User;
import org.springframework.jdbc.core.BeanPropertyRowMapper;
import org.springframework.jdbc.core.JdbcTemplate;
import org.springframework.jdbc.core.PreparedStatementCreator;
import org.springframework.jdbc.core.RowMapper;
import org.springframework.jdbc.support.GeneratedKeyHolder;
import org.springframework.jdbc.support.KeyHolder;
import org.springframework.stereotype.Repository;
import org.springframework.transaction.annotation.Transactional;
/**
 * @Repository 注解:标注这是一个持久化操作对象
 */
@Repository
public class UserRepository {
    // 注入 JdbcTemplate 模板对象
    @Resource
    private JdbcTemplate jdbcTemplate;

    /***
     * 插入数据
     * @return 插入影响的行数
     */
    public int insertUser(){
        String sql = "insert into tb_user(login_name ,username ,password) "
            + "values (?,?,?),(?,?,?),(?,?,?)";
        Object[] args = new Object[]{"swk","孙悟空","123456","zbj","猪八戒","123456","ts","唐僧","123456"};
        // 参数一:插入数据的 SQL 语句  参数二:对应 SQL 语句中占位符?的参数
        return jdbcTemplate.update(sql, args);
    }

    /***
```

```java
 * 根据 userName 查询数据
 * @param userName
 * @return User 对象
 */
public User selectByUsername(String username) {
    // 定义 SQL 语句
    String sql = "select * from tb_user where username = ?";
    // 定义一个 RowMapper
    RowMapper<User> rowMapper = new BeanPropertyRowMapper<>(User.class);
    // 执行查询方法
    User user = jdbcTemplate.queryForObject(sql, new Object[] { username }, rowMapper);
    return user;
}

/***
 * 根据 id 查询数据
 * @return User 对象
 */
public User findUserById(int id) {
    // 定义 SQL 语句
    String sql = "select * from tb_user where id=?";
    RowMapper<User> rowMapper = new BeanPropertyRowMapper<>(User.class);
    // 执行查询方法
    return jdbcTemplate.queryForObject(sql, new Object[] { id }, rowMapper);
}

/***
 * 查询所有数据
 * @return 包含 User 对象的 List 集合
 */
public List<User> findAll() {
    // 定义 SQL 语句
    String sql = "select * from tb_user";
    // 声明结果集的映射 rowMapper，将结果集的数据映射成 User 对象数据
    RowMapper<User> rowMapper = new BeanPropertyRowMapper<>(User.class);
    return jdbcTemplate.query(sql, rowMapper);
}

/***
 * 根据 id 删除数据
 */
public void delete(final Integer id) {
    // 定义 SQL 语句
    String sql = "delete from tb_user where id=?";
    // 执行
    jdbcTemplate.update(sql, new Object[] { id });
}

/***
 * 修改数据
 */
public void update(final User user) {
    // 定义 SQL 语句
    String sql = "update tb_user set username=?, login_name=? where id=?";
    // 执行
    jdbcTemplate.update(sql,
        new Object[] { user.getUsername(), user.getLoginName(), user.getId() });
}
/**
 * 插入数据，获取被插入数据的主键
 */
public User insertGetKey(User user) {
```

```java
        // 1.声明插入的SQL语句
        String sql = "insert into tb_user(username,login_name,password) values(?,?,?)";
        // 2.定义插入数据后获取主键的对象
        KeyHolder holder = new GeneratedKeyHolder();
        jdbcTemplate.update(new PreparedStatementCreator() {
            @Override
            public PreparedStatement createPreparedStatement(Connection connection) throws SQLException {
                // 3.插入数据后，将被插入数据的主键返回
                PreparedStatement ps = connection.prepareStatement(sql,
                        Statement.RETURN_GENERATED_KEYS);
                ps.setString(1, user.getUsername());
                ps.setString(2, user.getLoginName());
                ps.setString(3, user.getPassword());
                return ps;
            }
        }, holder);
        // 4.获取被插入数据库的主键，注入到user对象
        int newUserId = holder.getKey().intValue();
        user.setId(newUserId);
        return user;
    }
}
```

UserRepository 类需要通过@Repository 注解声明这是一个数据访问层对象，这样在业务层就可以通过注解注入 UserRepository 对象了。在 UserRepository 类中最关键的是通过注解@Resource 将 JdbcTemplate 对象注入进来，后续就可以通过 JdbcTemplate 对象提供的方法对数据库进行相关的 CRUD 操作了，JdbcTemplate 对象是由 Spring Boot 项目的自动配置和自动创建 Bean 完成的，所以使用起来十分方便。

⑤ 定义业务层类。

程序清单：codes/04/springbootjdbctemplate/src/main/java/org/fkit/springbootjdbctemplate/service/UserService.java

```java
import java.util.List;
import javax.annotation.Resource;
import org.fkit.springbootjdbctemplate.bean.User;
import org.fkit.springbootjdbctemplate.repository.UserRepository;
import org.springframework.stereotype.Service;

@Service
public class UserService {

    // 注入 UserRepository
    @Resource
    private UserRepository userRepository;

    public int insertUser(){
        return userRepository.insertUser();
    }

    public User selectByUsername(String username){
        return userRepository.selectByUsername(username);
    }

    public List<User> findAll(){
        return userRepository.findAll();
    }

    public User insertGetKey(User user) {
        return userRepository.insertGetKey(user);
```

```java
    }
    public void update(User user) {
        userRepository.update(user);
    }
    public void delete(Integer id) {
        userRepository.delete(id);
    }
}
```

在业务层中需要注入数据访问层对象 UserRepository，在上述代码中是通过@Resources 注解将 UserRepository 接口对应的实现类对象注入进来的。

⑥ 定义控制器类。

在 org.fkit.springbootjdbctemplate.controller 包中新建一个控制器类，命名为 UserController，其代码如下。

程序清单：codes/04/springbootjdbctemplate/src/main/java/org/fkit/springbootjdbctemplate/controller/UserController.java

```java
import java.util.List;
import javax.annotation.Resource;
import org.fkit.springbootjdbctemplate.bean.User;
import org.fkit.springbootjdbctemplate.service.UserService;
import org.springframework.web.bind.annotation.RequestMapping;
import org.springframework.web.bind.annotation.RestController;
@RestController
@RequestMapping("/user")
public class UserController {
    @Resource
    private UserService userService;

    @RequestMapping("/insertUser")
    public String insertUser(){
        return "插入数据["+userService.insertUser()+"]条";
    }

    @RequestMapping("/insertGetKey")
    public User insertGetKey(User user) {
        return userService.insertGetKey(user);
    }

    @RequestMapping("/selectByUsername")
    public User selectByUsername(String username){
        return userService.selectByUsername(username);
    }

    @RequestMapping("/findAll")
    public List<User> findAll(){
        return userService.findAll();
    }

    @RequestMapping("/update")
    public void update(User user) {
        userService.update(user);
    }

    @RequestMapping("/delete")
    public void delete(Integer id) {
```

```
            userService.delete(id);
    }
}
```

⑦ 测试应用。

启动 MySQL 数据库，在数据库中创建名为 jdbctemplate 的数据库，并创建 tb_user 表，执行脚本如下。

```
CREATE DATABASE jdbctemplate;
USE jdbctemplate;
CREATE TABLE tb_user(
    id INT PRIMARY KEY AUTO_INCREMENT,
    login_name VARCHAR(23),
    PASSWORD VARCHAR(23),
    username VARCHAR(23)
);
```

然后在 org.fkit.springbootjdbctemplate 包下新建 App.java 启动类，App.java 和之前的项目一致，此处不再赘述。右击该类运行 main 方法。Spring Boot 项目启动后，在浏览器中输入 URL 来测试应用。

测试添加用户信息，在浏览器中输入如下地址。

`http://127.0.0.1:8080/user/insertUser`

请求会提交给 UserController 类的 insertUser 方法进行处理，如果该方法保存用户成功，即可返回字符串"插入数据[3]条"。查看数据库数据，如图 4.34 所示。

图 4.34　插入用户数据

测试添加用户返回被插入数据的主键，在浏览器中输入如下地址。

`http://127.0.0.1:8080/user/insertGetKey?loginName=xl&username=徐磊&password=123456`

请求会提交给 UserController 类的 insertGetKey 方法进行处理，如果该方法保存用户成功，即可返回被插入的用户对象，该对象的 id 即是主键，如图 4.35 所示。

图 4.35　插入用户数据返回主键

测试查询所有用户，在浏览器中输入：

`http://127.0.0.1:8080/user/findAll`

请求会提交给 UserController 类的 findAll 方法进行处理，该方法查询出所有的用户信息，如图 4.36 所示。

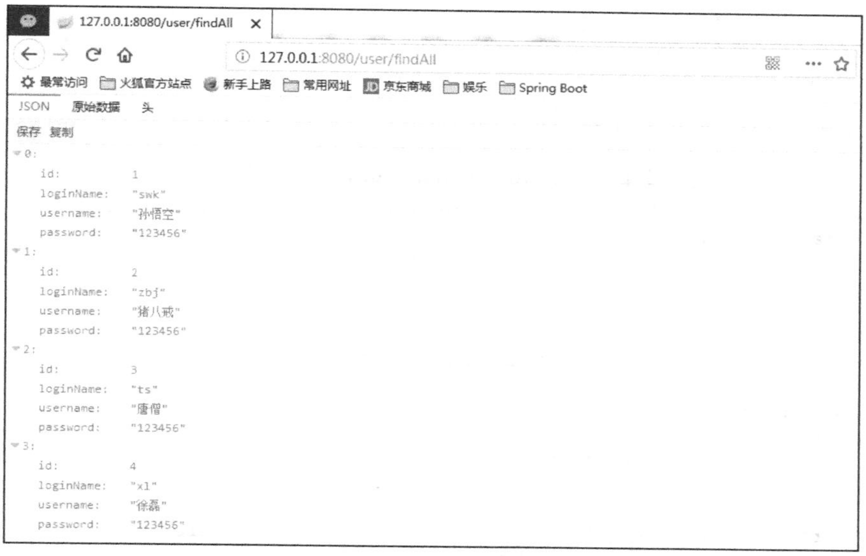

图 4.36 查询所有的用户信息

读者可自行更换条件进行更多的测试。

4.4 Spring Boot 整合 MyBatis

MyBatis 是一个数据持久层（ORM）框架。MyBatis 是优秀的执久层框架，在实体类和 SQL 语句之间建立了映射关系，MyBatis 支持普通 SQL 查询、存储过程和高级映射。MyBatis 消除了几乎所有的 JDBC 代码和参数的手工设置以及结果集的检索。MyBatis 使用简单的 XML 或注解进行配置和原始映射，将查询出的数据库的记录映射成 Java 对象。MyBatis 的开发可以在 Spring Boot 项目中进行整合。

示例：Spring Boot 整合 MyBatis 开发

创建一个新的 Maven 项目，命名为 springbootmybatistest。按照 Maven 项目的规范，在 src/main/ 下新建一个名为 resources 的文件夹。

① 修改 pom.xml 文件。

在 pom.xml 文件中增加 Web 开发依赖配置。

```xml
<dependency>
    <groupId>org.springframework.boot</groupId>
    <artifactId>spring-boot-starter-web</artifactId>
</dependency>
```

然后添加数据库驱动依赖配置。

```xml
<dependency>
    <groupId>mysql</groupId>
    <artifactId>mysql-connector-java</artifactId>
</dependency>
```

最后需要添加 mybatis-spring-boot-starter，它是 Spring Boot 项目整合 MyBatis 框架的核心依赖配置，该配置如下。

```xml
<dependency>
    <groupId>org.mybatis.spring.boot</groupId>
    <artifactId>mybatis-spring-boot-starter</artifactId>
```

```xml
        <!-- 1.3.1是作者成书时的最新版本 -->
        <version>1.3.1</version>
    </dependency>
```

修改后的完整 pom.xml 文件如下。

程序清单：codes/04/springbootmybatistest/pom.xml

```xml
<project xmlns="http://maven.apache.org/POM/4.0.0" xmlns:xsi="http://www.w3.org/2001/XMLSchema-instance"
    xsi:schemaLocation="http://maven.apache.org/POM/4.0.0 http://maven.apache.org/xsd/maven-4.0.0.xsd">
    <modelVersion>4.0.0</modelVersion>
    <groupId>org.fkit</groupId>
    <artifactId>springbootmybatistest</artifactId>
    <version>0.0.1-SNAPSHOT</version>
    <packaging>jar</packaging>
    <name>springbootmybatistest</name>
    <url>http://maven.apache.org</url>
    <parent>
        <groupId>org.springframework.boot</groupId>
        <artifactId>spring-boot-starter-parent</artifactId>
        <version>2.0.0.RELEASE</version>
    </parent>

    <properties>
        <project.build.sourceEncoding>UTF-8</project.build.sourceEncoding>
        <project.reporting.outputEncoding>UTF-8</project.reporting.outputEncoding>
        <java.version>1.8</java.version>
    </properties>
    <dependencies>
        <!-- 添加 spring-boot-starter-web 依赖... -->
        <dependency>
            <groupId>org.springframework.boot</groupId>
            <artifactId>spring-boot-starter-web</artifactId>
        </dependency>

        <!-- 添加 spring-boot-starter-thymeleaf 模块依赖 -->
        <dependency>
            <groupId>org.springframework.boot</groupId>
            <artifactId>spring-boot-starter-thymeleaf</artifactId>
        </dependency>

        <!-- 添加 MySQL 依赖 -->
        <dependency>
            <groupId>mysql</groupId>
            <artifactId>mysql-connector-java</artifactId>
        </dependency>

        <!-- 添加 MyBatis 依赖 -->
        <dependency>
            <groupId>org.mybatis.spring.boot</groupId>
            <artifactId>mybatis-spring-boot-starter</artifactId>
            <!-- 1.3.1是作者成书时的最新版本 -->
            <version>1.3.1</version>
        </dependency>
    </dependencies>
</project>
```

② 配置基本属性。

在 src/main/resources 包下新建一个全局配置文件，命名为 application.properties，在该配置文件中配置数据源。

程序清单：codes/04/springbootmybatistest/src/main/resources/application.properties
```
##########################################################
### 数据源信息配置
##########################################################
# 连接地址
spring.datasource.url=jdbc:mysql://localhost:3306/springbootmybatis
# 用户名配置
spring.datasource.username=root
# 密码
spring.datasource.password=
# 数据库驱动
spring.datasource.driver-class-name=com.mysql.jdbc.Driver
```

③ 创建持久化类。

在 org.fkit.springbootmybatistest 包下新建 4 个包，依次是 bean（放置持久化类）、controller（控制器）、repository（定义数据访问接口的包）、service（业务逻辑处理类），在 bean 包中创建一个持久化类 User.java，其代码如下。

程序清单：codes/04/springbootmybatistest/src/main/java/org/fkit/springbootmybatistest/bean/User.java
```java
import java.io.Serializable;
public class User implements Serializable {
    private int id;
    private String loginName;
    private String username;
    private String password;

    // 省略构造器和 set/get 方法……
}
```

④ 定义数据访问层接口。

在 org.fkit.springbootmybatistest.repository 包下新建一个接口，命名为 UserRepository.java，完整代码结构如下。

程序清单：codes/04/springbootmybatistest/src/main/java/org/fkit/springbootmybatistest/repository/UserRepository.java
```java
import java.util.List;
import org.apache.ibatis.annotations.Delete;
import org.apache.ibatis.annotations.Insert;
import org.apache.ibatis.annotations.Options;
import org.apache.ibatis.annotations.Param;
import org.apache.ibatis.annotations.Select;
import org.apache.ibatis.annotations.Update;
import org.fkit.springbootmybatistest.bean.User;
public interface UserRepository {

    @Insert("insert into tb_user(login_name , username , password) "
            + "values (#{loginName}, #{username}, #{ password })")
    public int insertUser(User user);

    /**
     * 插入数据获取主键
     */
    @Insert("insert into tb_user(login_name , username , password) "
            + "values (#{loginName}, #{username}, #{ password })")
    @Options(useGeneratedKeys=true, keyProperty="id", keyColumn="id")
    public void insertGetKey(User user);


    @Select("select * from tb_user where username = #{username}")
```

```java
    // 引用id="userResult"的@Results
    @ResultMap("userResult")
        public User selectByUsername(@Param("username")String username);

        @Select("select * from tb_user")
    // @Results用于映射对象属性和数据库列，常用于对象属性和数据库列不同名的情况
    @Results(id="userResult",value={
        @Result(id=true,column="id",property="id"),
        @Result(column="login_name",property="loginName"),
        @Result(column="password",property="password"),
        @Result(column="username",property="username")
    })
        public List<User> findAll();


        @Delete("delete from tb_user where id=#{id}")
        public void delete(final Integer id);

        @Select("select * from tb_user where id=#{id}")
    // 引用id="userResult"的@Results
    @ResultMap("userResult")
        public User findUserById(int id);

        @Update("update tb_user set username=#{username}, login_name=#{loginName} where id=#{id}")
        public void update(final User user);
}
```

⑤ 定义业务层类。

程序清单：codes/04/springbootmybatistest/src/main/java/org/fkit/springbootmybatistest/service/UserService.java

```java
import java.util.List;
import javax.annotation.Resource;
import org.fkit.springbootmybatistest.bean.User;
import org.fkit.springbootmybatistest.repository.UserRepository;
import org.springframework.stereotype.Service;
@Service
public class UserService {

    // 注入UserRepository
    @Resource
    private UserRepository userRepository;

    public int insertUser(User user){
        return userRepository.insertUser(user);
    }

    public User selectByUsername(String username){
        return userRepository.selectByUsername(username);
    }

    public List<User> findAll(){
        return userRepository.findAll();
    }

    public void insertGetKey(User user) {
        // 数据插入成功以后，MyBatis框架会将插入成功的数据主键存入user对象
        userRepository.insertGetKey(user);
    }

    public void update(User user) {
        userRepository.update(user);
    }
```

```java
    public void delete(Integer id) {
        userRepository.delete(id);
    }
}
```

⑥ 定义控制器类。

在 org.fkit.springbootmybatistest.controller 包中新建一个控制器类，命名为 UserController，完整代码如下。

程序清单：codes/04/springbootmybatistest/src/main/java/org/fkit/springbootmybatistest/controller/UserController.java

```java
import java.util.List;
import javax.annotation.Resource;
import org.fkit.springbootmybatistest.bean.User;
import org.fkit.springbootmybatistest.service.UserService;
import org.springframework.web.bind.annotation.RequestMapping;
import org.springframework.web.bind.annotation.RestController;
@RestController
@RequestMapping("/user")
public class UserController {

    // 注入UserService
    @Resource
    private UserService userService;

    @RequestMapping("/insertUser")
    public String insertUser(User user){
        return "插入数据["+userService.insertUser(user)+"]条";
    }

    @RequestMapping("/insertGetKey")
    public User insertGetKey(User user) {
        userService.insertGetKey(user);
        return user ;
    }

    @RequestMapping("/selectByUsername")
    public User selectByUsername(String username){
        return userService.selectByUsername(username);
    }

    @RequestMapping("/findAll")
    public List<User> findAll(){
        return userService.findAll();
    }

    @RequestMapping("/update")
    public void update(User user) {
        userService.update(user);
    }

    @RequestMapping("/delete")
    public void delete(Integer id) {
        userService.delete(id);
    }
}
```

⑦ 在 org.fkit.springbootmybatistest 包下新建 App.java 启动类，完整代码如下。

```java
import org.mybatis.spring.annotation.MapperScan;
import org.springframework.boot.SpringApplication;
```

```
import org.springframework.boot.autoconfigure.SpringBootApplication;
@SpringBootApplication
// 指定数据访问层接口的包名
@MapperScan("org.fkit.springbootmybatistest.repository")
public class App {
    public static void main(String[] args) {
        SpringApplication.run(App.class, args);
    }
}
```

根据 MyBatis 官方的说法，在 MyBatis 3 问世之前，Spring 3 的开发工作就已经完成了，所以 Spring 3（包括 Spring 4 和 Spring 5）中没有对 MyBatis 3 的支持。因此由 MyBatis 社区自己开发了一个 MyBatis-Spring 中间件用来满足 MyBatis 用户整合 Spring 的需求，该中间件有如下两个作用：

> 在 Spring 中配置 MyBatis 工厂类。
> 在持久层使用 Spring 注入的工具 Bean 对数据进行操作。

本书成书时该中间件的最高版本是 mybatis-spring-1.3.1.jar。该插件包可以在 Maven 依赖当中找到。@MapperScan 是在插件包中定义的注解，该注解的参数是一个包名字符串，只要在@MapperScan 中配置需要扫描的 MyBatis 接口的包路径，Spring 就会自动扫描包下面的 Java 接口，把这些类注册为 Spring 的 bean。

在本例当中，会自动扫描 org.fkit.springbootmybatistest.repository 包下的 UserRepository 接口，创建实例并注册为 Spring 的 bean。所以在服务层可以使用如下代码注入。

```
// 注入 UserRepository
@Resource
private UserRepository userRepository;
```

⑧ 测试应用。

启动 MySQL 数据库，在数据库中创建名为 springbootmybatis 的数据库，并创建 tb_user 表，执行脚本如下。

```
CREATE DATABASE springbootmybatis;
USE springbootmybatis;
CREATE TABLE tb_user(
    id INT PRIMARY KEY AUTO_INCREMENT,
    login_name VARCHAR(23),
    PASSWORD VARCHAR(23),
    username VARCHAR(23)
);
```

右击该类运行 main 方法。Spring Boot 项目启动后，在浏览器中输入 URL 来测试应用。
测试添加用户信息，在浏览器中输入如下信息。

```
http://127.0.0.1:8080/user/insertUser?loginName=swk&username=孙悟空&password=123456
```

请求会提交给 UserController 类的 insertUser 方法进行处理，如果该方法保存用户成功，即可返回字符串"插入数据[1]条"，查看数据库数据，如图 4.37 所示。

测试添加用户返回被插入数据的主键，在浏览器中输入如下信息。

```
http://127.0.0.1:8080/user/insertGetKey?loginName=xl&username=徐磊&password=123456
```

请求会提交给 UserController 类的 insertGetKey 方法进行处理，如果该方法保存用户成功，即可返回被插入的用户对象，该对象的 id 即是主键，如图 4.38 所示。

图 4.37 插入用户数据　　　　　图 4.38 插入用户数据返回主键

测试查询所有用户，在浏览器中输入如下信息。

```
http://127.0.0.1:8080/user/findAll
```

请求会提交给 UserController 类的 findAll 方法进行处理，该方法查询所有数据，如图 4.39 所示。

图 4.39 查询所有用户

读者可自行更换条件进行更多的测试。

4.5 本章小结

本章重点介绍了 Spring Boot 的数据访问开发，包括核心接口的介绍、Spring Data 的介绍和 Spring Data JPA 的介绍。本章以大量实用的示例进行了相关知识的介绍，如分页、动态查询以及各种查询方案等，同时讲解了整合 JdbcTemplate 和 MyBatis 的开发案例。

第 5 章将重点介绍 Spring Boot 的热部署与单元测试。

CHAPTER 5

第 5 章
Spring Boot 的热部署与单元测试

本章要点

- 使用 spring-boot-devtools 进行热部署
- Spring Boot 的单元测试

在项目开发过程中,常常会改动页面数据或者修改数据结构。为了显示改动效果,往往需要重启应用查看改变的效果,否则将不能看到新增代码的效果,这一过程很多时候是非常浪费时间的,导致开发效率极低。开发热部署可以在改变程序代码的时候,自动实现项目的重新启动和部署,大大提高了开发调试的效率。

5.1 使用 spring-boot-devtools 进行热部署

springloaded 并不能实现这些修改的热部署。spring-boot-devtools 是一个为开发者服务的模块,其中最重要的功能就是自动实现把更新的应用代码更改到最新的 App 上。其工作原理是在发现代码有更改之后,自动重新启动应用,但是速度比手动停止后再启动要更快。其深层原理是使用了两个 ClassLoader,一个 ClassLoader 加载那些不会改变的类(例如第三方的 Jar 包依赖),另一个 ClassLoader 加载会更改的类,称为 Restart ClassLoader。这样在有代码更改的时候,原来的 Restart ClassLoader 被丢弃,重新创建一个 Restart ClassLoader 加载更新的类,由于需要加载的类相对而言比较少,所以实现了较快的重启。

示例:使用 spring-boot-devtools 实现热部署

创建一个新的 Maven 项目,命名为 devtoolstest。

① 修改 pom.xml 文件。

在 pom.xml 文件中增加 Web 开发的启动器配置。

```xml
<dependency>
    <groupId>org.springframework.boot</groupId>
    <artifactId>spring-boot-starter-web</artifactId>
</dependency>
```

然后添加 spring-boot-devtools 的依赖配置。

```xml
<dependency>
    <groupId>org.springframework.boot</groupId>
    <artifactId>spring-boot-devtools</artifactId>
    <optional>true</optional>
    <scope>true</scope>
</dependency>
```

注意,还需要加入 spring-boot-maven-plugin。

```xml
<build>
    <plugins>
        <plugin>
            <groupId>org.springframework.boot</groupId>
            <artifactId>spring-boot-maven-plugin</artifactId>
            <configuration>
                <!-- 如果没有该项配置,devtools 不会起作用,即应用不会 restart -->
                <fork>true</fork>
            </configuration>
        </plugin>
    </plugins>
</build>
```

修改后的完整 pom.xml 文件如下。

程序清单:codes/05/devtoolstest/pom.xml

```xml
<project xmlns="http://maven.apache.org/POM/4.0.0" xmlns:xsi="http://www.w3.org/
2001/XMLSchema-instance"
    xsi:schemaLocation="http://maven.apache.org/POM/4.0.0
```

```xml
        http://maven.apache.org/xsd/maven-4.0.0.xsd">
    <modelVersion>4.0.0</modelVersion>

    <groupId>org.fkit</groupId>
    <artifactId>devtoolstest</artifactId>
    <version>0.0.1-SNAPSHOT</version>
    <packaging>jar</packaging>
    <name>devtoolstest</name>
    <url>http://maven.apache.org</url>
    <parent>
        <groupId>org.springframework.boot</groupId>
        <artifactId>spring-boot-starter-parent</artifactId>
        <version>2.0.0.RELEASE</version>
    </parent>

    <properties>
        <project.build.sourceEncoding>UTF-8</project.build.sourceEncoding>
        <project.reporting.outputEncoding>UTF-8</project.reporting.outputEncoding>
        <java.version>1.8</java.version>
    </properties>
    <dependencies>
        <dependency>
            <groupId>org.springframework.boot</groupId>
            <artifactId>spring-boot-starter-web</artifactId>
        </dependency>
        <!-- 添加 spring-boot-starter-thymeleaf 模块依赖 -->
        <dependency>
            <groupId>org.springframework.boot</groupId>
            <artifactId>spring-boot-starter-thymeleaf</artifactId>
        </dependency>

        <!-- Spring Boot spring-boot-devtools 依赖 -->
        <dependency>
            <groupId>org.springframework.boot</groupId>
            <artifactId>spring-boot-devtools</artifactId>
            <optional>true</optional>
            <scope>true</scope>
        </dependency>
    </dependencies>
    <!-- 添加 spring-boot-maven-plugin -->
    <build>
        <plugins>
            <plugin>
                <groupId>org.springframework.boot</groupId>
                <artifactId>spring-boot-maven-plugin</artifactId>
                <configuration>
                    <!-- 如果没有该项配置，devtools 不会起作用，即应用不会 restart -->
                    <fork>true</fork>
                </configuration>
            </plugin>
        </plugins>
    </build>
</project>
```

② 加入控制器类以及启动类。

在 org.fkit.devtoolstest 包下新建一个 controller 包，在其中定义一个控制器，命名为 HelloController.java，详细代码如下。

程序清单：codes/05/devtoolstest/src/main/java/org/fkit/devtoolstest/controller/HelloController.java

```java
package org.fkit.devtoolstest.controller;
import org.springframework.web.bind.annotation.RequestMapping;
```

```
import org.springframework.web.bind.annotation.RestController;
@RestController
public class HelloController {

    @RequestMapping("/hello")
    public String hello(){
        System.out.println("测试 spring-boot-devtools 热部署");
        return "hello,devtools……!";
    }
}
```

同时，在 org.fkit.devtoolstest 包下定义一个启动类，命名为 App.java，此处不再赘述。

③ 运行方式。

spring-boot-devtools 的项目无须配置启动参数，直接打开 App.java，右击鼠标，选择"Run As"下的"Java Application"即可。启动成功后在浏览器的输入框中输入如下地址。

http://127.0.0.1:8080/hello

请求触发 HelloController 类中的 hello 方法，返回结果如图 5.1 所示。

图 5.1　请求 hello 方法（一）

现在将 HelloController.java 类中的 hello 方法进行修改，修改后的代码如下：

```
package org.fkit.devtoolstest.controller;
import org.springframework.web.bind.annotation.RequestMapping;
import org.springframework.web.bind.annotation.RestController;
@RestController
public class HelloController {

    @RequestMapping("/hello")
    public String hello(){
        System.out.println("测试 spring-boot-devtools 热部署 2");
        return "hello2,devtools new……!";
    }
}
```

无须重启项目，直接在浏览器的输入框中输入如下地址：

http://127.0.0.1:8080/hello

请求触发 HelloController 类中的 hello 方法，控制台将输出"测试 spring-boot-devtools 热部署 2"，浏览器的返回结果如图 5.2 所示。

图 5.2　请求 hello 方法（二）

从此处可以看出，spring-boot-devtools 已经实现了方法体代码修改后的热部署，同样，spring-boot-devtools 也可以实现新增类、修改配置文件等的热部署。例如在 org.fkit.devtoolstest.controller 下新增一个控制器，命名为 HelloController2.java，代码如下。

程序清单：codes/05/devtoolstest/src/main/java/org/fkit/devtoolstest/controller/HelloController2.java

```java
package org.fkit.devtoolstest.controller;
import org.springframework.web.bind.annotation.RequestMapping;
import org.springframework.web.bind.annotation.RestController;
@RestController
public class HelloController2 {

    @RequestMapping("/hellonew")
    public String hello(){
        System.out.println("测试spring-boot-devtools new 热部署");
        return "hellonew,devtools new……!";
    }

}
```

无须重启项目，直接在浏览器的输入框中输入如下地址：

`http://127.0.0.1:8080/hellonew`

请求触发 HelloController2 类中的 hello 方法，浏览器返回的结果如图 5.3 所示。

图 5.3　请求 hello 方法（三）

又如新建 src/main/resources 目录，在该目录下创建一个名为 application.properties 的文件，配置内容如下。

程序清单：codes/05/devtoolstest/src/main/resources/application.properties

`server.port=9999`

无须重启项目，可以观察到控制器自动重启了应用，直接在浏览器的输入框中输入如下地址。

`http://127.0.0.1:9999/hellonew`

请求触发 HelloController2 类中的 hello 方法，浏览器返回的结果如图 5.4 所示。

图 5.4　请求 hello 方法（四）

可以观察到，现在应用的端口已经是 9999 了，配置文件的修改实现了热部署。spring-boot-devtools 还可以实现页面热部署（即页面修改后会立即生效，这个可以通过直接在 application.properties 文件中配置 spring.thymeleaf.cache=false 来实现（注意，不同的模板配置不一样）。这种方式由于自动实现了应用的重启，所以基本上我们所做的操作都可以实现热部署。

开发中如果出现了 spring-boot-devtools 不能实现热部署，一般可能是以下原因引起的：
- 对应的 spring-boot 版本是否正确，这里使用的是 2.0.0 版本。
- 是否加入 plugin 以及属性<fork>true</fork>。
- Eclipse Project 是否开启了 Build Automatically 自动编译的功能。
- 如果设置 SpringApplication.setRegisterShutdownHook(false)，则自动重启将不起作用。

5.2 Spring Boot 的单元测试

项目测试在系统开发中有着十分重要的作用,在大型系统,尤其是业务相对复杂的项目中,测试用例就显得尤为重要,它可以避免测试点的遗漏,提高测试效率,实现自动测试,在项目打包前进行测试校验,通过测试及时发现因为修改代码导致的新问题并及时解决。本节将重点介绍 Spring Boot 中的项目测试,示例采用了成书时最稳定的 Spring Boot 版本 2.0.0.RELEASE。在新版的 Spring Boot 中,官方推荐使用@SpringBootTest 定义测试类。

示例:使用 Spring Boot 的单元测试

创建一个新的 Maven 项目,命名为 springboottest。

① 修改 pom.xml 文件。

本示例的 Spring Boot 版本为 2.0.0.RELEASE,在 pom.xml 文件中增加 spring-boot-starter-test 启动器,本启动器作为 Spring Boot 单元测试的核心配置,具体配置如下。

```xml
<dependency>
    <groupId>org.springframework.boot</groupId>
    <artifactId>spring-boot-starter-test</artifactId>
    <scope>test</scope>
</dependency>
```

然后添加数据库驱动依赖配置,本案例是基于 MySQL 数据库操作的。

```xml
<dependency>
    <groupId>mysql</groupId>
    <artifactId>mysql-connector-java</artifactId>
</dependency>
```

最后需要添加 spring-boot-starter-data-jpa,它是 Spring Boot 项目访问数据库的核心依赖配置,加入此配置后,系统会自动导入 Spring Data 相关的核心数据访问接口包以及 Hibernate 框架相关的依赖包,该配置如下。

```xml
<dependency>
    <groupId>org.springframework.boot</groupId>
    <artifactId>spring-boot-starter-data-jpa</artifactId>
</dependency>
```

修改后的完整 pom.xml 文件如下。

程序清单:codes/05/springboottest/pom.xml

```xml
<project xmlns="http://maven.apache.org/POM/4.0.0" xmlns:xsi="http://www.w3.org/2001/XMLSchema-instance"
    xsi:schemaLocation="http://maven.apache.org/POM/4.0.0 http://maven.apache.org/ xsd/maven-4.0.0.xsd">
    <modelVersion>4.0.0</modelVersion>

    <groupId>org.fkit</groupId>
    <artifactId>springboottest</artifactId>
    <version>0.0.1-SNAPSHOT</version>
    <packaging>jar</packaging>
    <name>springboottest</name>
    <url>http://maven.apache.org</url>
    <parent>
        <groupId>org.springframework.boot</groupId>
        <artifactId>spring-boot-starter-parent</artifactId>
        <version>2.0.0.RELEASE</version>
    </parent>
    <properties>
        <project.build.sourceEncoding>UTF-8</project.build.sourceEncoding>
```

```xml
            <project.reporting.outputEncoding>UTF-8</project.reporting.outputEncoding>
            <java.version>1.8</java.version>
    </properties>
    <dependencies>

            <!-- 添加 spring-boot-starter-web 依赖 -->
            <dependency>
                <groupId>org.springframework.boot</groupId>
                <artifactId>spring-boot-starter-web</artifactId>
            </dependency>

<!-- 添加 spring-boot-starter-thymeleaf 模块依赖 -->
    <dependency>
            <groupId>org.springframework.boot</groupId>
            <artifactId>spring-boot-starter-thymeleaf</artifactId>
    </dependency>

            <!-- spring-boot-starter-test 依赖.... -->
            <dependency>
                <groupId>org.springframework.boot</groupId>
                <artifactId>spring-boot-starter-test</artifactId>
                <scope>test</scope>
            </dependency>

            <!-- 添加 MySQL 依赖 -->
            <dependency>
                <groupId>mysql</groupId>
                <artifactId>mysql-connector-java</artifactId>
            </dependency>

            <!-- 添加 Spring Data JPA 依赖 -->
            <dependency>
                <groupId>org.springframework.boot</groupId>
                <artifactId>spring-boot-starter-data-jpa</artifactId>
            </dependency>
    </dependencies>
</project>
```

② 配置基本属性。

在 src/main/resources 包下新建一个全局配置文件，命名为 application.properties，在该配置文件中配置数据源和 jpa 相关的属性。

程序清单：codes/05/springboottest/src/main/resources/application.properties

```
###########################################################
### 数据源信息配置
###########################################################
# 数据库地址
spring.datasource.url = jdbc:mysql://localhost:3306/springboottest
# 用户名
spring.datasource.username = root
# 密码
spring.datasource.password =
# 数据库驱动
spring.datasource.driverClassName = com.mysql.jdbc.Driver
# 指定连接池中最大的活跃连接数
spring.datasource.max-active=20
# 指定连接池最大的空闲连接数
spring.datasource.max-idle=8
# 指定必须保持连接的最小值
spring.datasource.min-idle=8
# 指定启动连接池时，初始建立的连接数量
```

```
spring.datasource.initial-size=10


##########################################################
### JPA 持久化配置
##########################################################
# 指定数据库的类型
spring.jpa.database = MySQL
# 指定是否需要在日志中显示 SQL 语句
spring.jpa.show-sql = true
# 指定自动创建|更新|验证数据库表结构等配置，配置成 update
# 表示如果数据库中存在持久化类对应的表就不创建，不存在就创建对应的表
spring.jpa.hibernate.ddl-auto = update
# 指定命名策略
spring.jpa.hibernate.naming-strategy = org.hibernate.cfg.ImprovedNamingStrategy
# 指定数据库方言
spring.jpa.properties.hibernate.dialect = org.hibernate.dialect.MySQL5Dialect
```

③ 创建持久化类。

在 org.fkit.springboottest.bean 包下新建 4 个包，分别是 bean（放置持久化类）、controller（控制器）、repository（定义数据访问接口的包）、service（业务逻辑处理类），在 bean 包中创建一个持久化类 Student.java，其代码如下。

程序清单：codes/05/springboottest/src/main/java/org/fkit/springboottest/bean/Student.java

```java
package org.fkit.springboottest.bean;
import java.io.Serializable;
import javax.persistence.Entity;
import javax.persistence.GeneratedValue;
import javax.persistence.GenerationType;
import javax.persistence.Id;
import javax.persistence.Table;

@Entity
@Table(name="tb_student")
public class Student implements Serializable{
    @Id
    @GeneratedValue(strategy = GenerationType.AUTO)
    private int id;
    private String name ;
    private String address ;
    private int age ;
    private char sex;
    public Student() {
        
    }
    public Student(String name, String address, int age, char sex) {
        super();
        this.name = name;
        this.address = address;
        this.age = age;
        this.sex = sex;
    }
    public int getId() {
        return id;
    }
    public void setId(int id) {
        this.id = id;
    }
    public String getName() {
        return name;
    }
    public void setName(String name) {
```

```java
        this.name = name;
    }
    public String getAddress() {
        return address;
    }
    public void setAddress(String address) {
        this.address = address;
    }
    public int getAge() {
        return age;
    }
    public void setAge(int age) {
        this.age = age;
    }
    public char getSex() {
        return sex;
    }
    public void setSex(char sex) {
        this.sex = sex;
    }
}
```

④ 定义数据访问层接口。

在 org.fkit.springboottest.repository 包下新建一个接口，命名为 StudentRepository，完整代码结构如下。

程序清单：codes/05/springboottest/src/main/java/org/fkit/springboottest/repository/StudentRepository.java

```java
package org.fkit.springboottest.repository;
import org.fkit.springboottest.bean.Student;
import org.springframework.data.jpa.repository.JpaRepository;

public interface StudentRepository extends JpaRepository<Student, Integer> {

}
```

⑤ 定义业务层类。

在 org.fkit.springboottest.service 包下新建一个业务层对象，命名为 SchoolService.java，代码如下。

程序清单：codes/05/springboottest/src/main/java/org/fkit/springboottest/service/SchoolService.java

```java
package org.fkit.springboottest.service;
import java.util.List;
import java.util.Map;
import javax.annotation.Resource;
import org.fkit.springboottest.bean.Student;
import org.fkit.springboottest.repository.StudentRepository;
import org.springframework.stereotype.Service;
import org.springframework.transaction.annotation.Transactional;
@Service
public class SchoolService {

    // 定义数据访问层接口对象
    @Resource
    private StudentRepository studentRepository;

    @Transactional
    public void save(Student stu) {
        studentRepository.save(stu);
    }

    public Student selectByKey(Integer id) {
```

```
        return studentRepository.findOne(id);
    }
}
```

⑥ 定义控制器类及启动类。

在 org.fkit.springboottest.controller 包中新建一个控制器类，命名为 StudentController，其代码如下。

程序清单：codes/05/springboottest/src/main/java/org/fkit/springboottest/controller/StudentController.java

```java
package org.fkit.springboottest.controller;
import java.util.HashMap;
import java.util.Map;
import javax.annotation.Resource;
import org.fkit.springboottest.bean.Student;
import org.fkit.springboottest.service.SchoolService;
import org.springframework.web.bind.annotation.PathVariable;
import org.springframework.web.bind.annotation.RequestBody;
import org.springframework.web.bind.annotation.RequestMethod;
import org.springframework.web.bind.annotation.ResponseBody;
import org.springframework.web.bind.annotation.RestController;
import org.springframework.web.bind.annotation.GetMapping;
import org.springframework.web.bind.annotation.PostMapping;

@RestController
@RequestMapping("/student")
public class StudentController {
    @Resource
    private SchoolService shcoolService;

    @PostMapping(value="/save")
    public Map<String,Object> save(@RequestBody Student stu) {
        shcoolService.save(stu);
        Map<String,Object> params = new HashMap<>();
        params.put("code", "success");
        return params;
    }

    /**
     * 获取学生信息
     * @param id
     */
    @GetMapping(value="/get/{id}")
    @ResponseBody
    public Student qryStu(@PathVariable(value = "id") Integer id){
        Student stu = shcoolService.selectByKey(id);
        return stu;
    }

}
```

在 org.fkit.springboottest 包下定义启动类 App.java，此处不再赘述。

⑦ Controller 单元测试。

Maven 项目为单元测试专门定义了一个包：src/test/java。在 src/test/java 的 org.fkit.springboottest 包下新建一个类，命名为 StudentControllerTest.java。需要注意的是，Spring Boot 项目中的测试类，需要在类上定义@RunWith(SpringRunner.class)，让测试运行于 Spring 的测试环境，以及注解@SpringBootTest(classes = App.class)，其中 App.class 是项目的启动类，其代码结构如下所示。

程序清单：codes/05/springboottest/src/test/java/org/fkit/springboottest/StudentControllerTest.java

```java
package org.fkit.springboottest;
import javax.transaction.Transactional;
import org.junit.Before;
import org.junit.Test;
import org.junit.runner.RunWith;
import org.springframework.beans.factory.annotation.Autowired;
import org.springframework.boot.test.context.SpringBootTest;
import org.springframework.http.MediaType;
import org.springframework.test.context.junit4.SpringRunner;
import org.springframework.test.web.servlet.MockMvc;
import org.springframework.test.web.servlet.request.MockMvcRequestBuilders;
import org.springframework.test.web.servlet.result.MockMvcResultHandlers;
import org.springframework.test.web.servlet.result.MockMvcResultMatchers;
import org.springframework.test.web.servlet.setup.MockMvcBuilders;
import org.springframework.web.context.WebApplicationContext;

@RunWith(SpringRunner.class)
@SpringBootTest(classes = App.class)
public class StudentControllerTest {

    // 注入Spring容器
    @Autowired
    private WebApplicationContext wac;
    // MockMvc 实现了对Http请求的模拟
    private MockMvc mvc;
    @Before
    public void setupMockMvc(){
        //初始化MockMvc对象
        mvc = MockMvcBuilders.webAppContextSetup(wac).build();
    }

    /**
     * 新增学生测试用例
     * @throws Exception
     */
    @Test
    @Transactional
    public void addStudent() throws Exception{
        String json="{\"name\":\"孙悟空\",\"address\":\"花果山\",\"age\":\"700\",\"sex\":\"男\"}";
        mvc.perform(MockMvcRequestBuilders.post("/student/save")
                .contentType(MediaType.APPLICATION_JSON_UTF8)
                .accept(MediaType.APPLICATION_JSON_UTF8)
                .content(json.getBytes())  //传递json参数
            )
            .andExpect(MockMvcResultMatchers.status().isOk())
            .andDo(MockMvcResultHandlers.print());
    }

    /**
     * 获取学生信息测试用例
     * @throws Exception
     */
    @Test
    public void qryStudent() throws Exception {
        mvc.perform(MockMvcRequestBuilders.get("/student/get/1")
                .contentType(MediaType.APPLICATION_JSON_UTF8)
                .accept(MediaType.APPLICATION_JSON_UTF8)
            )
            .andExpect(MockMvcResultMatchers.status().isOk())
```

```
            .andExpect(MockMvcResultMatchers.jsonPath("$.name").value("孙悟空"))
            .andExpect(MockMvcResultMatchers.jsonPath("$.address").value("花果山"))
            .andDo(MockMvcResultHandlers.print());
    }
}
```

在对 Controller 层（API）做测试时就得用到 MockMvc 了，可以不必启动工程就能测试这些接口。MockMvc 实现了对 HTTP 请求的模拟，能够直接使用网络的形式，转换到 Controller 的调用，这样可以使得测试速度快、不依赖网络环境，而且提供了一套验证的工具，这样可以使得请求的验证统一而且很方便。接下来对 StudentControllerTest 类中用到的 API 进行解释分析：

- mockMvc.perform 执行一个请求，请求可以调用到控制器中的某个方法。
- MockMvcRequestBuilders.get("/student/get/1")构造一个请求，Post 请求就用.post 方法。
- contentType(MediaType.APPLICATION_JSON_UTF8)代表发送端发送的数据格式是 application/json;charset=UTF-8。
- accept(MediaType.APPLICATION_JSON_UTF8)代表客户端希望接收的数据格式为 application/json;charset=UTF-8。
- ResultActions.andExpect 添加执行完成后的断言。
- ResultActions.andExpect(MockMvcResultMatchers.status().isOk())方法查看请求的状态响应码是否为 200，如果不是，则抛出异常，测试不通过。
- andExpect(MockMvcResultMatchers.jsonPath("$.name").value("孙悟空"))这里的 jsonPath 用来获取 name 字段，比对是否为孙悟空，不是就测试不通过。
- ResultActions.andDo 添加一个结果处理器，表示要对结果做点什么，比如此处使用 MockMvcResultHandlers.print()输出整个响应结果信息。

⑧ Controller 单元测试的执行。

首先需要启动 MySQL 数据库，在数据库中创建名为 springboottest 的数据库，脚本如下：

```
CREATE DATABASE springboottest;
```

打开 StudentControllerTest.java 测试类，选中 addStudent 方法，单击鼠标右键，选择 "Run As" → "JUnit Test" 命令，执行结果如图 5.5 所示。

图 5.5 单元测试（一）

切换到 "Console" 窗口，观察控制台，显示信息如下。

```
Hibernate: insert into tb_student (address, age, name, sex) values (?, ?, ?, ?)
MockHttpServletRequest:
      HTTP Method = POST
      Request URI = /student/save
       Parameters = {}
          Headers = {Content-Type=[application/json;charset=UTF-8], Accept=[application/json;charset=UTF-8]}
```

```
    MockHttpServletResponse:
              Status = 200
       Error message = null
             Headers = {Content-Type=[application/json;charset=UTF-8]}
        Content type = application/json;charset=UTF-8
                Body = {"code":"success"}
       Forwarded URL = null
      Redirected URL = null
             Cookies = []
```

可以看到测试成功，返回一个 json 数据{"code":"success"}，同时数据库的 tb_student 表插入一条数据。

打开 StudentControllerTest.java 测试类，选中 qryStudent 方法，单击鼠标右键，选择"Run As"→"JUnit Test"命令，执行结果如图 5.6 所示。

图 5.6　单元测试（二）

切换到"Console"窗口，观察控制台，显示信息如下：

```
    Hibernate: select student0_.id as id1_0_0_, student0_.address as address2_0_0_,
student0_.age as age3_0_0_, student0_.name as name4_0_0_, student0_.sex as sex5_0_0_
from tb_student student0_ where student0_.id=?
    MockHttpServletRequest:
         HTTP Method = GET
         Request URI = /student/get/1
          Parameters = {}
             Headers = {Content-Type=[application/json;charset=UTF-8], Accept=[application/
json;charset=UTF-8]}
    MockHttpServletResponse:
              Status = 200
       Error message = null
             Headers = {Content-Type=[application/json;charset=UTF-8]}
        Content type = application/json;charset=UTF-8
                Body = {"id":1,"name":"孙悟空","address":"花果山","age":700,"sex":"男"}
       Forwarded URL = null
      Redirected URL = null
             Cookies = []
```

可以看到测试成功，查询 tb_student 表 id 为 1 的记录，返回一个 json 数据{"id":1,"name":"孙悟空","address":"花果山","age":700,"sex":"男"}。

⑨ Service 单元测试。

在 src/test/java 的 org.fkit.springboottest 下定义业务层测试类，命名为 StudentServiceTest，代码如下。

程序清单：codes/05/springboottest/src/test/java/org/fkit/springboottest/StudentServiceTest.java

```
package org.fkit.springboottest;
import org.fkit.springboottest.bean.Student;
import org.fkit.springboottest.service.SchoolService;
import org.hamcrest.CoreMatchers;
```

```java
import org.junit.Assert;
import org.junit.Test;
import org.junit.runner.RunWith;
import org.springframework.beans.factory.annotation.Autowired;
import org.springframework.boot.test.context.SpringBootTest;
import org.springframework.test.context.junit4.SpringRunner;

@RunWith(SpringRunner.class)
@SpringBootTest
public class StudentServiceTest {
    @Autowired
    private SchoolService studentService;

    @Test
    public void findOne() throws Exception {
        Student stu = studentService.selectByKey(1);
        Assert.assertThat(stu .getName(),CoreMatchers.is("孙悟空"));
    }
}
```

打开 StudentServiceTest.java 测试类，选中 findOne 方法，单击鼠标右键，选择"Run As"→"JUnit Test"命令，执行结果如图 5.7 所示。

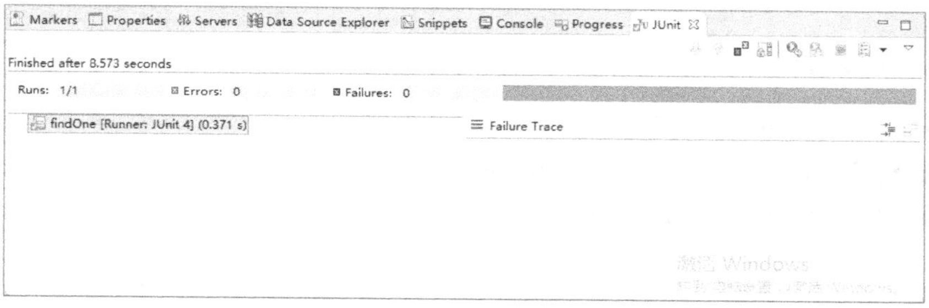

图 5.7 单元测试（三）

切换到"Console"窗口，观察控制台，显示信息如下。

```
Hibernate: select student0_.id as id1_0_0_, student0_.address as address2_0_0_, student0_.age as age3_0_0_, student0_.name as name4_0_0_, student0_.sex as sex5_0_0_ from tb_student student0_ where student0_.id=?
```

可以看到，发生了一个 SQL 语句查询 id 为 1 的记录，测试通过。

5.3 本章小结

本章主要介绍了 Spring Boot 的开发热部署，包括使用 spring-boot-devtools 实现热部署，并对 devtools 的特性进行了深入的分析；同时也介绍了 Spring Boot 项目中的测试，进行了案例操作与分析。

第 6 章将重点介绍 Spring Boot 的 Security 安全控制。

CHAPTER 6

第 6 章
Spring Boot 的 Security 安全控制

本章要点

- Spring Security 概念
- Spring Boot 对 Spring Security 的支持
- 企业 Spring Security 操作

Spring Security 是一个强大且高度可定制的身份验证和访问控制框架，完全基于 Spring 的应用程序的标准，Spring Security 为基于 Java EE 的企业应用程序提供了一个全面的安全解决方案。

6.1 Spring Security 是什么

Spring Security 是一个能够为基于 Spring 的企业应用系统提供安全访问控制解决方案的安全框架。它提供了一组可以在 Spring 应用上下文中配置的 Bean，充分利用了 Spring IoC（控制反转）和 AOP（面向切面编程）功能，为应用系统提供安全访问控制功能，减少了为企业系统安全控制编写大量重复代码的工作。

安全框架主要包括两个操作。

- 认证（Authentication）：确认用户可以访问当前系统。
- 授权（Authorization）：确定用户在当前系统中是否能够执行某个操作，即用户所拥有的功能权限。

Spring Security 包括多个模块。

- 核心模块（spring-security-core.jar）：包含核心的验证和访问控制类以及接口、远程支持和基本的配置 API。任何使用 Spring Security 的应用程序都需要这个模块。支持独立应用程序、远程客户端、服务层方法安全和 JDBC 用户配置。
- 远程调用（spring-security-remoting.jar）：提供与 Spring Remoting 的集成。
- Web 网页（spring-security-web.jar）：包括网站安全相关的基础代码，包括 Spring Security 网页验证服务和基于 URL 的访问控制。
- 配置（spring-security-config.jar）：包含安全命令空间的解析代码，如果使用 Spring Security XML 命令空间进行配置，需要使用该模块。
- LDAP（spring-security-ldap.jar）：LDAP 验证和配置代码，如果使用 LDAP 验证和管理 LDAP 用户实体，需要使用该模块。
- ACL 访问控制表（spring-security-acl.jar）：ACL 专门的领域对象的实现，用于在应用程序中对特定域对象实例应用安全性。
- CAS（spring-security-cas.jar）：Spring Security 的 CAS 客户端集成，用于 CAS 的 SSO 服务器使用 Spring Security 网页验证。
- OpenID（spring-security-openid.jar）：OpenID 网页验证支持，使用外部的 OpenID 服务器验证用户。
- Test（spring-security-test.jar）：支持 Spring Security 的测试。

在早期的 Spring Security 版本中使用 Spring Security 需要配置大量的 XML，本书所采用的最新版本全部基于 annotation 注解来完成 Spring Security 的功能。

6.2 Spring Security 入门

▶▶ 6.2.1 Security 适配器

在 Spring Boot 当中配置 Spring Security 非常简单，创建一个自定义类继承 WebSecurity-ConfigurerAdapter，并在该类中使用@EnableWebSecurity 注解，就可以通过重写 config 方法来配置所需要的安全配置。

WebSecurityConfigurerAdapter 是 Spring Security 为 Web 应用提供的一个适配器，实现了 WebSecurityConfigurer 接口，提供了两个方法用于重写开发者需要的安全配置。

```
protected void configure(HttpSecurity httpSecurity) throws Exception {
}
protected void configure(AuthenticationManagerBuilder auth) throws Exception {
}
```

configure(HttpSecurity httpSecurity)方法中可以通过 HttpSecurity 的 authorizeRequests() 方法定义哪些 URL 需要被保护、哪些不需要被保护；通过 formLogin()方法定义当需要用户登录的时候，跳转到的登录页面。

configureGlobal(AuthenticationManagerBuilder auth) 方法用于创建用户和用户的角色。

▶▶ 6.2.2 用户认证

Spring Security 是通过在 configureGlobal(AuthenticationManagerBuilder auth) 完成用户认证的。使用 AuthenticationManagerBuilder 的 inMemoryAuthentication()方法可以添加用户，并给用户指定权限。

```
@Autowired
    public void configureGlobal(AuthenticationManagerBuilder auth) throws Exception {
        auth.inMemoryAuthentication().withUser("fkit").password("123456").roles("USER");
        auth.inMemoryAuthentication().withUser("admin").password("admin").roles("ADMIN","DBA");
    }
```

上面的代码中添加了两个用户，其中一个用户名是"fkit"，密码是"123456"，用户权限是"USER"；另一个用户名是"admin"，密码是"admin"，用户权限有两个，分别是"ADMIN"和"DBA"。需要注意的是，Spring Security 保存用户权限的时候，会默认使用"ROLE_"，也就是说，"USER"实际上是"ROLE_USER"，"ADMIN"实际上是"ROLE_ADMIN"，"DBA"实际上是"ROLE_DBA"。

当然，也可以查询数据库获取用户和权限，常用的有 JPA、MyBatis 和 JDBC 操作，这些将在 6.3 节重点介绍。

▶▶ 6.2.3 用户授权

Spring Security 是通过 configure(HttpSecurity http) 完成用户授权的。

HttpSecurity 的 authorizeRequests()方法有多个子节点，每个 macher 按照它们的声明顺序执行，指定用户可以访问的多个 URL 模式。

➢ antMatchers 使用 Ant 风格匹配路径。
➢ regexMatchers 使用正则表达式匹配路径。

在匹配了请求路径后，可以针对当前用户的信息对请求路径进行安全处理。表 6.1 是 Spring Security 提供的安全处理方法。

表 6.1 安全处理方法

方法	用途
anyRequest	匹配所有请求路径
access(String)	Spring EL 表达式结果为 true 时可以访问
anonymous()	匿名可以访问
denyAll()	用户不能访问
fullyAuthenticated()	用户完全认证可以访问（非 remember-me 下自动登录）

续表

方法	用途
hasAnyAuthority(String…)	如果有参数，参数表示权限，则其中任何一个权限可以访问
hasAnyRole(String…)	如果有参数，参数表示角色，则其中任何一个角色可以访问
hasAuthority(String…)	如果有参数，参数表示权限，则其权限可以访问
hasIpAddress(String)	如果有参数，参数表示 IP 地址，如果用户 IP 和参数匹配，则可以访问
hasRole(String)	如果有参数，参数表示角色，则其角色可以访问
permitAll()	用户可以任意访问
rememberMe()	允许通过 remember-me 登录的用户访问
authenticated()	用户登录后可访问

示例代码如下：

```
@Override
protected void configure(HttpSecurity http) throws Exception {
    http.authorizeRequests()
    .antMatchers("/login").permitAll()
    .antMatchers("/", "/home").hasRole("USER")
    .antMatchers("/admin/**").hasAnyRole("ADMIN", "DBA")
    .anyRequest().authenticated();
}
```

以上代码解释如下。

> http.authorizeRequests()：开始请求权限配置。
> antMatchers("/login").permitAll()：请求匹配"/login"，所有用户都可以访问。
> antMatchers("/", "/home").hasRole("USER")：请求匹配"/"和"/home"，拥有"ROLE_USER"角色的用户可以访问。
> antMatchers("/admin/**"). hasAnyRole("ADMIN", "DBA")：请求匹配"/admin/**"，拥有"ROLE_ADMIN"或"ROLE_DBA"角色的用户可以访问。
> anyRequest().authenticated()：其余所有的请求都需要认证（用户登录）之后才可以访问。

HttpSecurity 还可以设置登录的行为，示例代码如下：

```
@Override
protected void configure(HttpSecurity http) throws Exception {
    http.authorizeRequests()
      .antMatchers("/login").permitAll()
      .antMatchers("/", "/home").hasRole("USER")
      .antMatchers("/admin/**").hasAnyRole("ADMIN", "DBA")
      .anyRequest().authenticated()
      .and()
      .formLogin()
    .loginPage("/login")
      .usernameParameter("loginName").passwordParameter("password")
      .defaultSuccessUrl("/success")
      .failureUrl("/login?error")
      .and()
      .logout()
    .permitAll()
      .and()
      .exceptionHandling().accessDeniedPage("/accessDenied");
}
```

以上代码的解释如下。

> formLogin()：开始设置登录操作。
> loginPage("/login")：设置登录页面的访问地址。
> usernameParameter("loginName").passwordParameter("password")：登录时接收传递的参

数"loginName"的值作为用户名，接收传递参数的"password"的值作为密码。
- defaultSuccessUrl("/success")：指定登录成功后转向的页面。
- failureUrl("/login?error")：指定登录失败后转向的页面和传递的参数。
- logout()：设置注销操作。
- permitAll()：所有用户都可以访问。
- exceptionHandling().accessDeniedPage("/accessDenied")：指定异常处理页面。

6.2.4 Spring Security 核心类

Spring Security 核心类包括 Authentication、SecurityContextHolder、UserDetails、UserDetailsService、GrantedAuthority、DaoAuthenticationProvider 和 PasswordEncoder。只要掌握了这些 Spring Security 的核心类，Spring Security 就会变得非常简单。

（1）Authentication

Authentication 用来表示用户认证信息，在用户登录认证之前，Spring Security 会将相关信息封装为一个 Authentication 具体实现类的对象，在登录认证成功之后又会生成一个信息更全面、包含用户权限等信息的 Authentication 对象，然后把它保存在 SecurityContextHolder 所持有的 SecurityContext 中，供后续的程序进行调用，如访问权限的鉴定等。

（2）SecurityContextHolder

SecurityContextHolder 是用来保存 SecurityContext 的。SecurityContext 中含有当前所访问系统的用户的详细信息。默认情况下，SecurityContextHolder 将使用 ThreadLocal 来保存 SecurityContext，这也就意味着在处于同一线程的方法中，可以从 ThreadLocal 获取到当前的 SecurityContext。

Spring Security 使用一个 Authentication 对象来描述当前用户的相关信息。SecurityContextHolder 中持有的是当前用户的 SecurityContext，而 SecurityContext 持有的是代表当前用户相关信息的 Authentication 的引用。这个 Authentication 对象不需要我们自己创建，在与系统交互的过程中，Spring Security 会自动创建相应的 Authentication 对象，然后赋值给当前的 SecurityContext。开发过程中常常需要在程序中获取当前用户的相关信息，比如最常见的是获取当前登录用户的用户名。示例代码如下：

```
String username = SecurityContextHolder.getContext().getAuthentication().getName();
```

（3）UserDetails

UserDetails 是 Spring Security 的一个核心接口。其中定义了一些可以获取用户名、密码、权限等与认证相关的信息的方法。Spring Security 内部使用的 UserDetails 实现类大都是内置的 User 类，要使用 UserDetails，也可以直接使用该类。在 Spring Security 内部，很多需要使用用户信息的时候，基本上都是使用 UserDetails，比如在登录认证的时候。

通常需要在应用中获取当前用户的其他信息，如 E-mail、电话等。这时存放在 Authentication 中的 principal 只包含认证相关信息的 UserDetails 对象可能就不能满足我们的要求了。这时可以实现自己的 UserDetails，在该实现类中可以定义一些获取用户其他信息的方法，这样将来就可以直接从当前 SecurityContext 的 Authentication 的 principal 中获取这些信息。

UserDetails 是通过 UserDetailsService 的 loadUserByUsername() 方法进行加载的。UserDetailsService 也是一个接口，我们也需要实现自己的 UserDetailsService 来加载自定义的 UserDetails 信息。

（4）UserDetailsService

Authentication.getPrincipal() 的返回类型是 Object，但很多情况下返回的其实是一个 UserDetails 的实例。登录认证的时候 Spring Security 会通过 UserDetailsService 的 loadUserByUsername() 方法获取对应的 UserDetails 进行认证，认证通过后会将该 UserDetails 赋给认证通过的 Authentication 的 principal，然后再把该 Authentication 存入 SecurityContext。之后如果需要使用用户信息，可以通过 SecurityContextHolder 获取存放在 SecurityContext 中的 Authentication 的 principal。

（5）GrantedAuthority

Authentication 的 getAuthorities() 可以返回当前 Authentication 对象拥有的权限，即当前用户拥有的权限。其返回值是一个 GrantedAuthority 类型的数组，每一个 GrantedAuthority 对象代表赋予给当前用户的一种权限。GrantedAuthority 是一个接口，其通常是通过 UserDetailsService 进行加载，然后赋予 UserDetails 的。

GrantedAuthority 中只定义了一个 getAuthority() 方法，该方法返回一个字符串，表示对应的权限，如果对应权限不能用字符串表示，则应当返回 null。

（6）DaoAuthenticationProvider

Spring Security 默认会使用 DaoAuthenticationProvider 实现 AuthenticationProvider 接口，专门进行用户认证的处理。DaoAuthenticationProvider 在进行认证的时候需要一个 UserDetailsService 来获取用户的信息 UserDetails，其中包括用户名、密码和所拥有的权限等。如果需要改变认证的方式，开发者可以实现自己的 AuthenticationProvider。

（7）PasswordEncoder

在 Spring Security 中，对密码的加密都是由 PasswordEncoder 来完成的。在 Spring Security 中，已经对 PasswordEncoder 有了很多实现，包括 md5 加密、SHA-256 加密等，开发者只要直接拿来用就可以。在 DaoAuthenticationProvider 中，有一个就是 PasswordEncoder 属性，密码加密功能主要靠它来完成。

在 Spring 的官方文档中明确指出，如果开发一个新的项目，BCryptPasswordEncoder 是较好的选择。BCryptPasswordEncoder 使用 BCrypt 的强散列哈希加密实现，并可以由客户端指定加密的强度，强度越高安全性自然就越高。本书示例正是使用 BCryptPasswordEncoder 进行加密的。

▶▶ 6.2.5 Spring Security 的验证机制

Spring Security 大体上是由一堆 Filter 实现的，Filter 会在 Spring MVC 前拦截请求。Filter 包括登出 Filter（LogoutFilter）、用户名密码验证 Filter（UsernamePasswordAuthenticationFilter）之类。Filter 再交由其他组件完成细分的功能，最常用的 UsernamePasswordAuthenticationFilter 会持有一个 AuthenticationManager 引用，AuthenticationManager 是一个验证管理器，专门负责验证。但 AuthenticationManager 本身并不做具体的验证工作，AuthenticationManager 持有一个 AuthenticationProvider 集合，AuthenticationProvider 才是做验证工作的组件，验证成功或失败之后调用对应的 Handler。

▶▶ 6.2.6 Spring Boot 的支持

Spring Boot 针对 Spring Security 提供了自动配置的功能，这些默认的自动配置极大地简化了开发工作。

（1）Spring Boot 通过 org.springframework.boot.autoconfigure.security 包对 Spring Security

提供了自动配置的支持，其工作主要通过 SecurityAutoConfiguration 和 SecurityProperties 两个类来完成自动配置。

SecurityProperties 类使用以"security"为前缀的属性配置 Spring Security 相关的配置，具体内容如下：

```
security.user.name=user      # 默认用户名
security.user.password=      # 默认用户密码
security.user.role=USER      # 默认用户角色
security.require-ssl=false   # 是否需要 ssl 支持
security.enable-csrf=false   # 是否开启 csrf 支持，默认关闭
# 默认 basic 认证设置
security.basic.enabled=true
security.basic.realm=Spring
security.basic.path= # /**
# 默认 headers 认证设置
security.headers.xss=false
security.headers.cache=false
security.headers.frame=false
security.headers.contentType=false
security.headers.hsts=all
security.sessions=stateless  # Session 创建策略(always, never, if_required, stateless)
security.ignored=false # 安全策略
```

Spring Boot 已经为开发者做了如此多的配置，当需要自己扩展时，只需要自定义类继承 WebSecurityConfigurerAdapter，无须再使用@EnableWebSecurity 注解。

（2）Spring Boot 自动配置一个 DefaultSecurityFilterChain 过滤器，用来忽略/css/**、/js/**、/images/**、/webjars/**、/**/favicon.ico、/error 等文件的拦截。

（3）Spring Boot 自动注册 Security 的过滤器。

示例：简单 Spring Boot Security 应用

接下来通过一个简单的 Spring Boot 应用，演示在 Spring Boot 中 Spring Security 的使用。创建一个新的 Maven 项目，命名为 securitytest。按照 Maven 项目的规范，在 src/main/下新建一个名为 resources 的文件夹，并在 src/main/resources 下新建 static 和 templates 两个文件夹。

① 修改 pom.xml 文件，并引入静态文件。

pom.xml 文件和之前大致相同，只是本例使用了 Spring Security，需要引入相关模块。

程序清单：codes/06/securitytest/pom.xml

```xml
<!-- 添加 spring-boot-starter-security 模块 -->
<dependency>
    <groupId>org.springframework.boot</groupId>
    <artifactId>spring-boot-starter-security</artifactId>
</dependency>
```

② 开发用于测试的 html 页面。

程序清单：codes/06/securitytest/src/main/resources/templates/login.html

```html
<!DOCTYPE html>
<html xmlns="http://www.w3.org/1999/xhtml"
    xmlns:th="http://www.thymeleaf.org"
    xmlns:sec="http://www.thymeleaf.org/thymeleaf-extras-springsecurity4">
<head>
    <title>Spring Boot Security 示例</title>
    <link rel="stylesheet" th:href="@{css/bootstrap.min.css}" />
    <link rel="stylesheet" th:href="@{css/app.css}" />
    <link rel="stylesheet" th:href="@{css/bootstrap-theme.min.css}"/>
```

```html
    <link rel="stylesheet" type="text/css" href="//cdnjs.cloudflare.com/ajax/libs/font-awesome/4.2.0/css/font-awesome.css" />
    <script type="text/javascript" th:src="@{js/jquery-1.11.0.min.js}"></script>
    <script type="text/javascript" th:src="@{js/bootstrap.min.js}"></script>
    <script type="text/javascript">
        $(function(){
            $("#loginBtn").click(function(){
                var loginName = $("#loginName");
                var password = $("#password");
                var msg = "";
                if(loginName.val() == ""){
                    msg = "登录名称不能为空!";
                    loginName.focus();
                }else if(password.val() == ""){
                    msg = "密码不能为空!";
                    password.focus();
                }
                if(msg != ""){
                    alert(msg);
                    return false;
                }
                $("#loginForm").submit();
            });
        });
    </script>
</head>
<body>
<div class="panel panel-primary">
    <div class="panel-heading">
        <h3 class="panel-title">简单 Spring Boot Security 示例</h3>
    </div>
</div>
<div id="mainWrapper">
    <div class="login-container">
        <div class="login-card">
            <div class="login-form">
                <!-- 表单提交到 login -->
                <form id="loginForm" th:action="@{/login}" method="post" class="form-horizontal">
                    <!-- 用户名或密码错误提示 -->
                    <div th:if="${param.error != null}">
                        <div class="alert alert-danger">
                            <p><font color="red">用户名或密码错误!</font></p>
                        </div>
                    </div>
                    <!-- 注销提示 -->
                    <div th:if="${param.logout != null}">
                        <div class="alert alert-success">
                            <p><font color="red">用户已注销成功!</font></p>
                        </div>
                    </div>
                    <div class="input-group input-sm">
                        <label class="input-group-addon" ><i class="fa fa-user"></i></label>
                        <input type="text" class="form-control" id="loginName" name="loginName" placeholder="请输入用户名" />
                    </div>
                    <div class="input-group input-sm">
                        <label class="input-group-addon" ><i class="fa fa-lock"></i></label>
                        <input type="password" class="form-control" id="password" name="password" placeholder="请输入密码" />
                    </div>
```

```html
                    <div class="form-actions">
                        <input id="loginBtn" type="button"
                            class="btn btn-block btn-primary btn-default" value="登录"/>
                    </div>
                </form>
            </div>
        </div>
    </div>
</body>
</html>
```

login.html 用于向 login 请求提交 loginname 和 password，从而进行登录。

程序清单：codes/06/securitytest/src/main/resources/templates/home.html

```html
<!DOCTYPE html>
<html xmlns:th="http://www.thymeleaf.org">
<head>
<meta charset="UTF-8"></meta>
<title>home 页面</title>
<link rel="stylesheet" th:href="@{css/bootstrap.min.css}" />
<link rel="stylesheet" th:href="@{css/bootstrap-theme.min.css}"/>
<script type="text/javascript" th:src="@{js/jquery-1.11.0.min.js}"></script>
<script type="text/javascript" th:src="@{js/bootstrap.min.js}"></script>
</head>
<body>
<div class="panel panel-primary">
    <div class="panel-heading">
        <h3 class="panel-title">Home 页面</h3>
    </div>
</div>
    <h3>欢迎[<font color="red"><span th:text="${user}">用户名</span></font>]访问 Home 页面!
    您的权限是<font color="red"><span th:text="${role}">权限</span></font><br/><br/>
    <a href="admin">访问 admin 页面</a><br/><br/>
    <a href="logout">安全退出</a></h3>
</body>
</html>
```

home.html 是 ROLE_USER 用户登录之后显示的页面，同时提供了一个超链接到 admin 页面。

程序清单：codes/06/securitytest/src/main/resources/templates/admin.html

```html
<!DOCTYPE html>
<html xmlns:th="http://www.thymeleaf.org">
<head>
<meta charset="UTF-8"></meta>
<title>admin 页面</title>
<link rel="stylesheet" th:href="@{css/bootstrap.min.css}" />
<link rel="stylesheet" th:href="@{css/bootstrap-theme.min.css}"/>
<script type="text/javascript" th:src="@{js/jquery-1.11.0.min.js}"></script>
<script type="text/javascript" th:src="@{js/bootstrap.min.js}"></script>
</head>
<body>
<div class="panel panel-primary">
    <div class="panel-heading">
        <h3 class="panel-title">Admin 页面</h3>
    </div>
</div>
    <h3>欢迎[<font color="red"><span th:text="${user}">用户名</span></font>]访问 Admin 页面!
```

```html
您的权限是<font color="red"><span th:text="${role}">权限</span></font><br/><br/>
<a href="dba">访问 dba 页面</a><br/><br/>
<a href="logout">安全退出</a></h3>
</body>
</html>
```

admin.html 是 ROLE_ADMIN 用户登录之后显示的页面,同时提供了一个到 dba 页面的超链接。

程序清单:codes/06/securitytest/src/main/resources/templates/dba.html

```html
<!DOCTYPE html>
<html xmlns:th="http://www.thymeleaf.org">
<head>
<meta charset="UTF-8"></meta>
<title>dba 页面</title>
<link rel="stylesheet" th:href="@{css/bootstrap.min.css}" />
<link rel="stylesheet" th:href="@{css/bootstrap-theme.min.css}"/>
<script type="text/javascript" th:src="@{js/jquery-1.11.0.min.js}"></script>
<script type="text/javascript" th:src="@{js/bootstrap.min.js}"></script>
</head>
<body>
<div class="panel panel-primary">
    <div class="panel-heading">
        <h3 class="panel-title">DBA 页面</h3>
    </div>
</div>
    <h3>欢迎[<font color="red"><span th:text="${user}">用户名</span></font>]访问 DBA 页面!
    您的权限是<font color="red"><span th:text="${role}">权限</span></font><br/><br/>
    <a href="logout">安全退出</a></h3>
</body>
</html>
```

dba.html 只是显示简单的欢迎语句。

程序清单:codes/06/securitytest/src/main/resources/templates/accessDenied.html

```html
<!DOCTYPE html>
<html xmlns:th="http://www.thymeleaf.org">
<head>
<meta charset="UTF-8"></meta>
<title>访问拒绝页面</title>
<link rel="stylesheet" th:href="@{css/bootstrap.min.css}" />
<link rel="stylesheet" th:href="@{css/bootstrap-theme.min.css}"/>
<script type="text/javascript" th:src="@{js/jquery-1.11.0.min.js}"></script>
<script type="text/javascript" th:src="@{js/bootstrap.min.js}"></script>
</head>
<body>
<div class="panel panel-primary">
    <div class="panel-heading">
        <h3 class="panel-title">AccessDenied 页面</h3>
    </div>
</div>
    <h3><font color="red"><span th:text="${user}">用户名</span></font>,您没有权限访问页面!
    您的权限是<font color="red"><span th:text="${role}">权限</span></font><br/> <br/>
    <a href="logout">安全退出</a></h3>
</body>
</html>
```

accessDenied.html 是访问拒绝页面,如果登录的用户没有权限访问该页面,会进行提示。

③ 创建 Spring Security 的认证处理类。

程序清单：codes/06/securitytest/src/main/java/org/fkit/securitytest/security/ MyPasswordEncoder.java

```java
import org.springframework.security.crypto.password.PasswordEncoder;
public class MyPasswordEncoder implements PasswordEncoder{
    @Override
    public String encode(CharSequence arg0) {
        return arg0.toString();
    }
    @Override
    public boolean matches(CharSequence arg0, String arg1) {
        return arg1.equals(arg0.toString());
    }
}
```

Spring Security 5.0 之后为了更加安全，修改了密码存储格式，密码存储格式为：{id}encodedPassword。id 是一个标识符，用于查找是哪个 PasswordEncoder，也就是密码加密的格式所对应的 PasswordEncoder。encodedPassword 是指原始加密经过加密之后的密码。id 必须在密码的开始，id 前后必须加{}。如果 id 找不到，id 则会为空，会抛出异常：There is no PasswordEncoder mapped for the id "null"。

设置密码的示例代码如下。

```
auth.inMemoryAuthentication().withUser("fkit").password("{noop}123456").roles("USER");
```

除了直接使用密码存储格式，Spring Security 还提供了密码编辑器接口 PasswordEncoder，在实际项目开发中,可以使用自定义密码编辑器,自定义密码编辑器类必须实现PasswordEncoder接口。

程序清单：codes/06/securitytest/src/main/java/org/fkit/securitytest/security/AppSecurityConfigurer.java

```java
import org.springframework.beans.factory.annotation.Autowired;
import org.springframework.context.annotation.Configuration;
import org.springframework.security.config.annotation.authentication.builders.AuthenticationManagerBuilder;
import org.springframework.security.config.annotation.web.builders.HttpSecurity;
import org.springframework.security.config.annotation.web.configuration.WebSecurityConfigurerAdapter;
@Configuration
public class AppSecurityConfigurer extends WebSecurityConfigurerAdapter{
    /**
     * 注入认证处理类，处理不同用户跳转到不同的页面
     */
    @Autowired
    AppAuthenticationSuccessHandler appAuthenticationSuccessHandler;
    /**
     * 用户授权操作
     */
    @Override
    protected void configure(HttpSecurity http) throws Exception {
        System.out.println("AppSecurityConfigurer configure() 调用......");
        http.authorizeRequests()
        //spring-security 5.0 之后需要过滤静态资源
            .antMatchers("/login", "/css/**","/js/**","/img/*").permitAll()
            .antMatchers("/", "/home").hasRole("USER")
            .antMatchers("/admin/**").hasAnyRole("ADMIN", "DBA")
            .anyRequest().authenticated()
            .and()
            .formLogin().loginPage("/login")
            .successHandler(appAuthenticationSuccessHandler)
```

```
                .usernameParameter("loginname").passwordParameter("password")
                .and()
                .logout().permitAll()
                .and()
                .exceptionHandling().accessDeniedPage("/accessDenied");
    }
    /**
     * 用户认证操作
     */
    @Autowired
    public void configureGlobal(AuthenticationManagerBuilder auth) throws Exception {
        System.out.println("AppSecurityConfigurer configureGlobal() 调用......");
        // spring-security 5.0 之后需要密码编码器,否则会抛出异常: There is no PasswordEncoder
mapped for the id "null"
        auth.inMemoryAuthentication().passwordEncoder(new MyPasswordEncoder()).
withUser("fkit").password("123456").roles("USER");
        auth.inMemoryAuthentication().passwordEncoder(new MyPasswordEncoder()).
withUser("admin").password("admin").roles("ADMIN","DBA");
    }
}
```

AppSecurityConfigurer 类是本示例最关键的一个类,用于处理 Spring Security 的用户认证和授权操作。注意加粗的两行代码:

```
@Autowired
AppAuthenticationSuccessHandler appAuthenticationSuccessHandler;
```

注入了一个认证处理类,登录成功之后不同用户需要跳转到不同的页面,所以此处使用了一个 AppAuthenticationSuccessHandler 类进行处理。

```
successHandler(appAuthenticationSuccessHandler)
```

successHandler(AuthenticationSuccessHandler successHandler)方法用于处理登录成功之后的操作。

config()和 configureGlobal()两个方法中分别打印了一句话,用于启动应用时的跟踪调试。

④ 创建认证成功处理类。

程序清单:codes/06/securitytest/src/main/java/org/fkit/securitytest/security/
AppAuthenticationSuccessHandler.java

```java
import java.io.IOException;
import java.util.ArrayList;
import java.util.Collection;
import java.util.List;
import javax.servlet.http.HttpServletRequest;
import javax.servlet.http.HttpServletResponse;
import org.springframework.security.core.Authentication;
import org.springframework.security.core.GrantedAuthority;
import org.springframework.security.web.DefaultRedirectStrategy;
import org.springframework.security.web.RedirectStrategy;
import org.springframework.security.web.authentication.SimpleUrlAuthentication-
SuccessHandler;
import org.springframework.stereotype.Component;
@Component
public class AppAuthenticationSuccessHandler extends SimpleUrlAuthentication-
SuccessHandler{
    
    // Spring Security 通过 RedirectStrategy 对象负责所有重定向事务
    private RedirectStrategy redirectStrategy = new DefaultRedirectStrategy();
    
    /*
     * 重写 handle 方法,方法中通过 RedirectStrategy 对象重定向到指定的 URL
     */
```

```java
    @Override
    protected void handle(HttpServletRequest request, HttpServletResponse response,
            Authentication authentication)
            throws IOException {
        // 通过determineTargetUrl方法返回需要跳转的URL
        String targetUrl = determineTargetUrl(authentication);
        // 重定向请求到指定的URL
        redirectStrategy.sendRedirect(request, response, targetUrl);
    }

    /*
     * 从Authentication对象中提取当前登录用户的角色，并根据其角色返回适当的URL
     */
    protected String determineTargetUrl(Authentication authentication) {
        String url = "";

        // 获取当前登录用户的角色权限集合
        Collection<? extends GrantedAuthority> authorities = authentication.getAuthorities();

        List<String> roles = new ArrayList<String>();

        // 将角色名称添加到List集合
        for (GrantedAuthority a : authorities) {
            roles.add(a.getAuthority());
        }

        // 判断不同角色跳转到不同的URL
        if (isAdmin(roles)) {
            url = "/admin";
        } else if (isUser(roles)) {
            url = "/home";
        } else {
            url = "/accessDenied";
        }
        System.out.println("url = " + url);
        return url;
    }

    private boolean isUser(List<String> roles) {
        if (roles.contains("ROLE_USER")) {
            return true;
        }
        return false;
    }

    private boolean isAdmin(List<String> roles) {
        if (roles.contains("ROLE_ADMIN")) {
            return true;
        }
        return false;
    }

    public void setRedirectStrategy(RedirectStrategy redirectStrategy) {
        this.redirectStrategy = redirectStrategy;
    }

    protected RedirectStrategy getRedirectStrategy() {
        return redirectStrategy;
    }
}
```

AppAuthenticationSuccessHandler类继承了SimpleUrlAuthenticationSuccessHandler类，该类

继承了 AbstractAuthenticationTargetUrlRequestHandler 父类，实现了 AuthenticationSuccessHandler 接口，是 Spring 用来处理用户认证授权并跳转到指定 URL 的。

重写 handler 方法，方法中通过获得当前用户需要跳转的 URL，并使用 Spring Security 负责重定向事务的 RedirectStrategy 对象重定向到指定的 URL。

⑤ 转发请求的控制器。

程序清单：codes/06/securitytest/src/main/java/org/fkit/securitytest/controller/AppController.java

```java
import javax.servlet.http.HttpServletRequest;
import javax.servlet.http.HttpServletResponse;
import org.springframework.security.core.Authentication;
import org.springframework.security.core.context.SecurityContextHolder;
import org.springframework.security.core.userdetails.UserDetails;
import org.springframework.security.web.authentication.logout.SecurityContextLogoutHandler;
import org.springframework.stereotype.Controller;
import org.springframework.ui.Model;
import org.springframework.web.bind.annotation.RequestMapping;
@Controller
public class AppController {

    @RequestMapping("/")
    public String index() {
        return "index";
    }

    @RequestMapping("/home")
    public String homePage(Model model) {
        model.addAttribute("user", getUsername());
        model.addAttribute("role", getAuthority());
        return "home";
    }

    @RequestMapping(value = "/login")
    public String login() {
        return "login";
    }

    @RequestMapping(value = "/admin")
    public String adminPage(Model model) {
        model.addAttribute("user", getUsername());
        model.addAttribute("role", getAuthority());
        return "admin";
    }

    @RequestMapping(value = "/dba")
    public String dbaPage(Model model) {
        model.addAttribute("user", getUsername());
        model.addAttribute("role", getAuthority());
        return "dba";
    }

    @RequestMapping(value = "/accessDenied")
    public String accessDeniedPage(Model model) {
        model.addAttribute("user", getUsername());
        model.addAttribute("role", getAuthority());
        return "accessDenied";
```

```java
    }
    @RequestMapping(value="/logout")
     public String logoutPage (HttpServletRequest request, HttpServletResponse response) {
        // Authentication 是一个接口，用来表示用户认证信息
        Authentication auth = SecurityContextHolder.getContext().getAuthentication();
        // 如果用户认证信息不为空，注销
        if (auth != null){
            new SecurityContextLogoutHandler().logout(request, response, auth);
        }
        // 重定向到 login 页面
        return "redirect:/login?logout";
    }

    /**
     * 获得当前用户名称
     * */
    private String getUsername(){
        String username = SecurityContextHolder.getContext().getAuthentication().getName();
        System.out.println("username = " + username);
        return username;
    }
    /**
     * 获得当前用户权限
     * */
    private String getAuthority(){
        // 获得 Authentication 对象，表示用户认证信息
        Authentication authentication = SecurityContextHolder.getContext().getAuthentication();
        List<String> roles = new ArrayList<String>();
        // 将角色名称添加到 List 集合
        for (GrantedAuthority a : authentication.getAuthorities()) {
            roles.add(a.getAuthority());
        }
        System.out.println("role = " + roles);
        return roles.toString();
    }

}
```

AppController 是一个 Spring MVC 的控制器，提供了响应 login、home、admin、dba、accessDenied 和 logout 请求的方法。每个方法通过 getUsername()方法获得当前认证用户的用户名，通过 getAuthority()方法获得当前认证用户的权限，并设置到 Model 当中。

在 getUsername()方法、getAuthority()方法和 logoutPage()方法中都使用了 Authentication 对象。Authentication 是一个接口，用来表示用户认证信息，在用户登录认证之前，相关信息会封装为一个 Authentication 具体实现类的对象，在登录认证成功之后又会生成一个信息更全面、包含用户权限等信息的 Authentication 对象，然后把它保存在 SecurityContextHolder 所持有的 SecurityContext 中，供后续的程序进行调用，如访问权限的鉴定等。

⑥ 测试应用。

App.java 和之前的项目一致，此处不再赘述。运行 main 方法启动项目。

Spring Boot 项目启动后，观察控制台，发现自定义类 AppSecurityConfigurer 的两个方法都已经被执行，说明自定义的用户认证和用户授权工作已经生效。

```
AppSecurityConfigurer configureGlobal() 调用......
AppSecurityConfigurer configure() 调用......
```

在浏览器中输入 URL 测试应用：

http://localhost "/"、"/login"、"home"、"admin" 等任何一个当前项目的请求都会被重定向到 http://localhost:8080/login 页面，因为没有登录，用户没有访问权限，如图 6.1 所示。

图 6.1 登录页面

输入任意登录名、密码，单击"登录"按钮。提示用户名或密码错误，如图 6.2 所示。

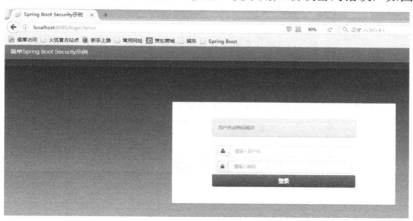

图 6.2 用户名或密码错误

输入用户名 fkit，密码 123456，单击"登录"按钮。该用户是"ROLE_USER"，跳转到 Home 页面，如图 6.3 所示。

图 6.3 访问 Home 页面

单击超链接"访问 admin 页面"，由于当前用户的权限只是"ROLE_USER"，不能访问 admin 页面，所以跳转到访问拒绝页面，如图 6.4 所示。

图 6.4 访问拒绝页面

单击超链接"安全退出",跳转到 logout 请求,控制器执行注销用户认证信息代码,再次跳转到登录页面,提示"用户已注销成功",如图 6.5 所示。

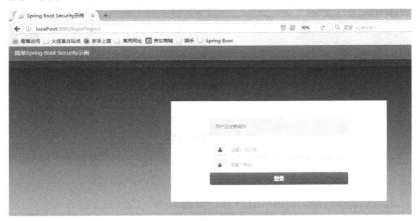

图 6.5 用户已注销成功

输入用户名 admin,密码 admin,单击"登录"按钮。该用户是"ROLE_ADMIN"和"ROLE_DBA",跳转到 admin 页面,如图 6.6 所示。

图 6.6 访问 admin 页面

单击超链接"访问 dba 页面",由于当前用户的权限是"ROLE_ADMIN"和"ROLE_DBA",可以访问 dba 页面,所以跳转到 dba 页面,如图 6.7 所示。

图 6.7 访问 dba 页面

通过测试可以看到,securitytest 项目已经使用 Spring Security 实现了用户认证和用户授权。

6.3 企业项目中的 Spring Security 操作

在企业真实项目中，通常将用户和权限信息保存在企业数据库当中，本节将重点介绍真实项目中的 Spring Security 操作，通过验证登录名 loginName 和密码 password 进行认证并授权，同时密码使用加密处理，避免将密码明文存储到数据库。操作企业数据库主流的方式有 JPA、MyBatis 和 JDBC，接下来逐个介绍。

示例：基于 JPA 的 Spring Boot Security 操作

创建一个新的 Maven 项目，命名为 securityjpatest。按照 Maven 项目的规范，在 src/main/下新建一个名为 resources 的文件夹，并在 src/main/resources 下新建 static 和 templates 两个文件夹。

① 修改 pom.xml 文件，并引入静态文件。

pom.xml 文件和之前大致相同，只是本示例使用了 JPA 操作数据库，需要引入相关模块。

程序清单：codes/06/securityjpatest/pom.xml

```xml
<!-- 添加 spring-boot-starter-data-jpa 模块 -->
<dependency>
    <groupId>org.springframework.boot</groupId>
    <artifactId>spring-boot-starter-data-jpa</artifactId>
</dependency>
<!-- 添加 MySQL 数据库驱动 -->
<dependency>
    <groupId>mysql</groupId>
    <artifactId>mysql-connector-java</artifactId>
</dependency>
```

② 在配置文件 application.properties 中加入数据库和 JPA 的相关配置信息。

程序清单：codes/06/securityjpatest/src/main/resources/application.properties

```
spring.datasource.driver-class-name=com.mysql.jdbc.Driver
spring.datasource.url=jdbc:mysql://localhost:3306/springboot
spring.datasource.username=root
spring.datasource.password=
logging.level.org.springframework.security=trace
spring.thymeleaf.cache=false
spring.jpa.hibernate.ddl-auto=update
spring.jpa.show-sql=true
```

③ 开发用于测试的 html 页面，方式和 6.2 节中的示例一致，包括 login.html、home.html、admin.html、dba.html 和 accessDenied.html。

④ 创建用户类和角色类。

程序清单：codes/06/securityjpatest/src/main/java/org/fkit/securityjpatest/pojo/FKRole.java

```java
import javax.persistence.Entity;
import javax.persistence.GeneratedValue;
import javax.persistence.GenerationType;
import javax.persistence.Id;
import javax.persistence.Table;
import java.io.Serializable;
import javax.persistence.Column;

@Entity
@Table(name="tb_role")
public class FKRole implements Serializable {
    @Id
```

```java
    @GeneratedValue(strategy=GenerationType.IDENTITY)
    @Column(name="id")
    private Long id;
    @Column(name="authority")
    private String authority;

    public FKRole() {
        super();
        // TODO Auto-generated constructor stub
    }
    public Long getId() {
        return id;
    }
    public void setId(Long id) {
        this.id = id;
    }

    public String getAuthority() {
        return authority;
    }
    public void setAuthority(String authority) {
        this.authority = authority;
    }
    @Override
    public String toString() {
        return "FKRole [id=" + id + ", authority=" + authority + "]";
    }
}
```

FKRole 类用来保存权限信息，有 id 和 authority 两个属性，authority 表示用户权限。

程序清单：codes/06/securityjpatest/src/main/java/org/fkit/securityjpatest/pojo/FKUser.java

```java
import java.io.Serializable;
import java.util.List;
import javax.persistence.CascadeType;
import javax.persistence.Column;
import javax.persistence.Entity;
import javax.persistence.FetchType;
import javax.persistence.GeneratedValue;
import javax.persistence.GenerationType;
import javax.persistence.Id;
import javax.persistence.JoinTable;
import javax.persistence.JoinColumn;
import javax.persistence.ManyToMany;
import javax.persistence.Table;

@Entity
@Table(name="tb_user")
public class FKUser implements Serializable{

    @Id
    @GeneratedValue(strategy=GenerationType.IDENTITY)
    @Column(name="id")
private Long id;
    private String loginName;
    private String username;
    private String password;

    @ManyToMany(cascade = {CascadeType.REFRESH},fetch = FetchType.EAGER)
    @JoinTable(name="tb_user_role",
        joinColumns={@JoinColumn(name="user_id")},
        inverseJoinColumns={@JoinColumn(name="role_id")})
    private List<FKRole> roles;
```

```java
    public Long getId() {
        return id;
    }
    public void setId(Long id) {
        this.id = id;
    }
    public String getLoginName() {
        return loginName;
    }
    public void setLoginName(String loginName) {
        this.loginName = loginName;
    }
    public String getUsername() {
        return username;
    }
    public void setUsername(String username) {
        this.username = username;
    }
    public String getPassword() {
        return password;
    }
    public void setPassword(String password) {
        this.password = password;
    }
    public List<FKRole> getRoles() {
        return roles;
    }
    public void setRoles(List<FKRole> roles) {
        this.roles = roles;
    }

    @Override
    public String toString() {
        return "User [id=" + id + ", loginName=" + loginName + ", username=" + username
            + ", password=" + password
            + ", roles=" + roles + "]";
    }
}
```

FKUser 类用来保存用户数据。为了更接近真实项目，这里定义了 5 个属性：id、loginName、password、username 和 roles。其中 loginName 是登录名，唯一；username 是用户真实姓名，登录时输入唯一的 loginName，欢迎页面显示欢迎用户真实姓名 username；roles 表示用户权限的 List 集合，用户和权限的关系是多对多关系。

> **注意**
> 实际开发中，FKUser 还可以实现 org.springframework.security.core.userdetails.UserDetails 接口，实现该接口之后 FKUser 即可成为 Spring Security 所使用的用户，本示例为了区分 JPA 的 pojo 和 Spring Security 的用户对象，并没有实现 UserDetails 接口，而是在 UserDetailsService 接口中进行绑定。

⑤ 创建数据访问接口。

程序清单：codes/06/securityjpatest/src/main/java/org/fkit/securityjpatest/repository/UserRepository.java

```java
import org.fkit.securityjpatest.pojo.FKUser;
import org.springframework.data.jpa.repository.JpaRepository;
public interface UserRepository extends JpaRepository<FKUser, Long>{

    // 根据登录名查询用户
    FKUser findByLoginName(String loginName);

}
```

自定义持久层类 UserRepository 实现了 JpaRepository 接口，注意 Spring-data-jpa 的规范，此处是根据登录名 loginName 进行查询，所以方法名是 findByLoginName。

⑥ 创建自定义服务类 Service。

程序清单：codes/06/securityjpatest/src/main/java/org/fkit/securityjpatest/service/UserService.java

```java
import java.util.ArrayList;
import java.util.List;
import org.fkit.securityjpatest.pojo.FKRole;
import org.fkit.securityjpatest.pojo.FKUser;
import org.fkit.securityjpatest.repository.UserRepository;
import org.springframework.beans.factory.annotation.Autowired;
import org.springframework.security.core.GrantedAuthority;
import org.springframework.security.core.authority.SimpleGrantedAuthority;
import org.springframework.security.core.userdetails.User;
import org.springframework.security.core.userdetails.UserDetails;
import org.springframework.security.core.userdetails.UserDetailsService;
import org.springframework.security.core.userdetails.UsernameNotFoundException;

/**
 * 需要实现 UserDetailsService 接口
 * 因为在 Spring Security 中配置的相关参数需要是 UserDetailsService 类型的数据
 * */
@Service
public class UserService implements UserDetailsService{

    // 注入持久层接口 UserRepository
    @Autowired
    UserRepository userRepository;

    /*
     * 重写 UserDetailsService 接口中的 loadUserByUsername 方法，通过该方法查询对应的用户 (non-Javadoc)
     * 返回对象 UserDetails 是 Spring Security 的一个核心接口。
     * 其中定义了一些可以获取用户名、密码、权限等与认证相关信息的方法
     */
    @Override
    public UserDetails loadUserByUsername(String username) throws UsernameNotFoundException {
        // 调用持久层接口 findByLoginName 方法查找用户，此处传进来的参数实际是 loginName
        FKUser fkUser = userRepository.findByLoginName(username);
        if (fkUser == null) {
            throw new UsernameNotFoundException("用户名不存在");
        }
        // 创建 List 集合，用来保存用户权限，GrantedAuthority 对象代表赋予当前用户的权限
        List<GrantedAuthority> authorities = new ArrayList<>();
        // 获得当前用户权限集合
        List<FKRole> roles = fkUser.getRoles();
        for (FKRole role : roles) {
            // 将关联对象 Role 的 authority 属性保存为用户的认证权限
```

```java
        authorities.add(new SimpleGrantedAuthority(role.getAuthority()));
    }
    // 此处返回的是 org.springframework.security.core.userdetails.User 类，该类是
    // Spring Security 内部的实现
    return new User(fkUser.getUsername(), fkUser.getPassword(), authorities);
}
```

自定义的 UserService 实现了 UserDetailsService 接口，登录认证的时候 Spring Security 会通过 UserDetailsService 的 loadUserByUsername()方法获取对应的 UserDetails 进行认证。

此处的重点是重写了 UserDetailsService 接口的 loadUserByUsername()方法，在方法中调用持久层接口的 findByLoginname()方法通过 JPA 进行数据库验证，传递的参数是页面接收到的 loginName。最后将获得的用户名、密码和权限保存到 org.springframework.security.core.userdetails.User 类中并返回，该 User 类是 Spring Security 内部的实现，专门用于保存用户名、密码、权限等与认证相关的信息。

⑦ 创建 Spring Security 的认证处理类。

程序清单：codes/06/securityjpatest/src/main/java/org/fkit/securityjpatest/security/AppSecurityConfigurer.java

```java
import org.fkit.securityjpatest.service.UserService;
import org.springframework.beans.factory.annotation.Autowired;
import org.springframework.context.annotation.Bean;
import org.springframework.context.annotation.Configuration;
import org.springframework.security.config.annotation.authentication.builders.AuthenticationManagerBuilder;
import org.springframework.security.config.annotation.web.builders.HttpSecurity;
import org.springframework.security.config.annotation.web.configuration.WebSecurityConfigurerAdapter;
import org.springframework.security.core.userdetails.UserDetailsService;

/**
 * 自定义 Spring Security 认证处理类的时候
 * 需要继承自 WebSecurityConfigurerAdapter，重写对应的方法来完成相关的认证处理
 */
@Configuration
public class AppSecurityConfigurer extends WebSecurityConfigurerAdapter{

    // 依赖注入用户服务类
    @Autowired
    private UserService userService;

    // 依赖注入加密接口
    @Autowired
    private PasswordEncoder passwordEncoder;

    // 依赖注入用户认证接口
    @Autowired
    private AuthenticationProvider authenticationProvider;

    // 依赖注入认证处理成功类，验证用户成功后处理不同用户跳转到不同的页面
    @Autowired
    AppAuthenticationSuccessHandler appAuthenticationSuccessHandler;

    /*
     * BCryptPasswordEncoder 是 Spring Security 提供的 PasswordEncoder 接口，是实现类
     * 用来创建密码的加密程序，避免将密码明文存储到数据库
     */
    @Bean
    public PasswordEncoder passwordEncoder() {
```

```java
        return new BCryptPasswordEncoder();
    }

    // DaoAuthenticationProvider 是 Spring Security 提供的 AuthenticationProvider 实现
    @Bean
    public AuthenticationProvider authenticationProvider() {
        // 创建 DaoAuthenticationProvider 对象
        DaoAuthenticationProvider provider = new DaoAuthenticationProvider();
        // 不要隐藏"用户未找到"异常
        provider.setHideUserNotFoundExceptions(false);
        // 通过重写 configure 方法添加自定义的认证方式。
        provider.setUserDetailsService(userService);
        // 设置密码加密程序认证
        provider.setPasswordEncoder(passwordEncoder);
        return provider;
    }

    @Override
    protected void configure(AuthenticationManagerBuilder auth) throws Exception {
        // 设置认证方式
        auth.authenticationProvider(authenticationProvider);
    }

    /**
     * 设置了登录页面,而且登录页面任何人都可以访问,然后设置了登录失败地址,也设置了注销请求,
     * 注销请求也是任何人都可以访问的。
     * permitAll 表示该请求任何人都可以访问, .anyRequest().authenticated()表示其他的请求
     * 都必须要有权限认证。
     */
    @Override
    protected void configure(HttpSecurity http) throws Exception {
        http.authorizeRequests()
        // spring-security 5.0 之后需要过滤静态资源
          .antMatchers("/login","/css/**","/js/**","/img/*").permitAll()
          .antMatchers("/", "/home").hasRole("USER")
          .antMatchers("/admin/**").hasAnyRole("ADMIN", "DBA")
          .anyRequest().authenticated()
          .and()
          .formLogin().loginPage("/login").successHandler(appAuthenticationSuccessHandler)
          .usernameParameter("loginName").passwordParameter("password")
          .and()
          .logout().permitAll()
          .and()
          .exceptionHandling().accessDeniedPage("/accessDenied");
    }
}
```

AppSecurityConfigurer 类继承 WebSecurityConfigurerAdapter 类,用于处理 Spring Security 的用户认证和授权操作。重点是加粗的代码:

```java
@Override
    protected void configure(AuthenticationManagerBuilder auth) throws Exception {
        // 添加自定义的认证方式
        auth.authenticationProvider(authenticationProvider);
    }
```

使用 AuthenticationManagerBuilder 的 authenticationProvider ()方法进行认证。

在 authenticationProvider ()方法中调用 setUserDetailsService ()方法传入自定义的 UserService 对象进行用户登录认证,Spring Security 会通过 UserDetailsService 的 loadUserByUsername()方法获取对应的 UserDetails 进行认证,认证通过后会将该 UserDetails 赋给认证通过的

Authentication 的 principal，然后再把该 Authentication 存入 SecurityContext。

在 authenticationProvider ()方法中调用 setPasswordEncoder()方法传入 Spring Security 提供的 BCryptPasswordEncoder 对象进行密码加密验证。

⑧ 测试应用。

授权操作、认证成功处理类、控制器等和之前的示例一致，此处不再赘述。

App.java 和之前的项目一致，此处也不再赘述。运行 main 方法启动项目。

根据对象之间的关系，JPA 会在数据库中自动创建 tb_user、tb_role 和中间表 tb_user_role 三张表。

需要注意的是，按照 JPA 的规范，实体类 FKUser 的属性 loginName 映射到数据库的时候，自动生成的数据库列名是 login_name。

往数据库插入测试数据，执行如下 SQL 语句：

```
INSERT INTO `tb_role`(`id`,`authority`) VALUES (1,'ROLE_ADMIN'),(2,'ROLE_DBA'),
(3,'ROLE_USER');
INSERT INTO `tb_user`(`id`,login_name,`password`,`username`)
VALUES (1,'admin','$2a$10$VgFGmXCoM0eJE45Ej4aOEeEFD6kvemnd2HFXhcMMherTQGZW0rZw.',
'管理员'),(2,'fkit','$2a$10$KRVhKjIvEQQ1ENWM0k.Iwe8Dr6HPdu56Rfn9qJ6FFrpPFJf6IaiN.',
'疯狂软件');
INSERT INTO `tb_user_role`(`user_id`,`role_id`) VALUES (1,1),(1,2),(2,3);
```

> **注意**
> 本示例在 tb_user 表中插入的 password 列的数据是加密后的密码，用户在页面登录时 admin 用户的密码是 admin，fkit 的密码是 123456。真实项目开发应该通过注册模块对用户密码进行加密，此处没有提供注册模块，所以直接插入已经加密过的密码进行测试。

读者也可以自己写一个 main 方法，通过 new BCryptPasswordEncoder().encode("admin")代码给自己的数据库密码进行加密，只要修改 encode()方法中的参数即可。

在浏览器中输入 URL 来测试应用。

```
http://localhost:8080
```

此时和之前的示例一样，跳转到登录页面，如图 6.1 所示。输入用户名 admin，密码 admin，单击"登录"按钮。该用户的权限是"ROLE_ADMIN"和"ROLE_DBA"，跳转到 admin 页面，如图 6.8 所示。

图 6.8 测试管理员登录

控制台显示的执行的 SQL 语句如下：

```
Hibernate: select fkuser0_.id as id1_1_, fkuser0_.loginname as loginnam2_1_,
fkuser0_.password as password3_1_, fkuser0_.username as username4_1_ from tb_user
fkuser0_ where fkuser0_.loginname=?
Hibernate: select roles0_.user_id as user_id1_2_0_, roles0_.role_id as
```

```
role_id2_2_0_, fkrole1_.id as id1_0_1_, fkrole1_.authority as authorit2_0_1_ from
tb_user_role roles0_ inner join tb_role fkrole1_ on roles0_.role_id=fkrole1_.id where
roles0_.user_id=?
```

第一句通过 loginName 查询了用户的信息，第二句通过用户 id 关联查询了用户的权限。

在登录页面输入的 loginName 是"admin"，页面显示的是用户的 username 的值"管理员"。

读者可以根据之前示例的步骤进行测试，测试结果完全一样。区别在于 6.2 节的示例是使用字符串进行用户认证授权，而本示例是通过查询数据库中的数据进行用户认证授权，在真实项目的开发过程中，实用性更强。

示例：基于 MyBatis 的 Spring Boot Security 操作

创建一个新的 Maven 项目，命名为 securitymybatistest。按照 Maven 项目的规范，在 src/main/下新建一个名为 resources 的文件夹，并在 src/main/resources 下新建 static 和 templates 两个文件夹。该项目和 securityjpatest 的所有页面、数据库表结构、表数据完全一致，只是修改了几处代码，重点是持久层使用的是 MyBatis 框架。

① 修改 pom.xml 文件，并引入静态文件。

pom.xml 文件和之前大致相同，只是本示例使用了 MyBatis 操作数据库，需要引入相关模块。

```xml
<!-- 添加 MyBatis 模块 -->
<dependency>
    <groupId>org.mybatis.spring.boot</groupId>
    <artifactId>mybatis-spring-boot-starter</artifactId>
    <version>1.3.1</version>
</dependency>
<!-- 添加 MySQL 数据库驱动 -->
<dependency>
    <groupId>mysql</groupId>
    <artifactId>mysql-connector-java</artifactId>
</dependency>
```

由于 Spring 官方没有支持 MyBatis 的相关依赖模块，MyBatis 官方增加了一个名为 mybatis-spring-boot-starter 的依赖模块，所以此处需要加上版本号，在本书成书之时的最高版本是 1.3.1。

② 在配置文件 application.properties 中加入数据库和 MyBatis 的相关配置信息。

程序清单：codes/06/securitymybatistest/src/main/resources/application.properties

```
spring.datasource.driver-class-name=com.mysql.jdbc.Driver
spring.datasource.url=jdbc:mysql://localhost:3306/springboot
spring.datasource.username=root
spring.datasource.password=
logging.level.org.springframework.security=info
# 可以在控制台观察输出的 SQL 语句
logging.level.org.fkit.securitymybatistest.mapper.UserMapper=debug
spring.thymeleaf.cache=false
```

Spring Boot 会自动加载 spring.datasource.* 相关配置，数据源就会自动注入 sqlSessionFactory，sqlSessionFactory 会自动注入 Mapper，开发者就可以直接使用了。

③ 创建数据访问接口。

程序清单：codes/06/securitymybatistest/src/main/java/org/fkit/securitymybatistest/mapper/UserMapper

```java
import java.util.List;
import org.apache.ibatis.annotations.Many;
import org.apache.ibatis.annotations.Result;
```

```java
import org.apache.ibatis.annotations.Results;
import org.apache.ibatis.annotations.Select;
import org.apache.ibatis.mapping.FetchType;
import org.fkit.securitymybatistest.pojo.FKRole;
import org.fkit.securitymybatistest.pojo.FKUser;
public interface UserMapper {

    // 根据loginName查询用户信息，同时关联查询用户的权限
    @Select("select * from tb_user where login_name = #{loginName}")
    @Results({
            @Result(id=true,column="id",property="id"),
            @Result(column="login_name",property="loginName"),
            @Result(column="password",property="password"),
            @Result(column="username",property="username"),
            @Result(column="id",property="roles",
            many=@Many(select="findRoleByUser",
            fetchType=FetchType.EAGER))
    })
    FKUser findByLoginName(String loginName);

    // 根据用户id关联查询用户的所有权限
    @Select(" SELECT id,authority FROM tb_role r,tb_user_role ur "
        + " WHERE r.id = ur.role_id AND user_id = #{id}")
    List<FKRole> findRoleByUser(Long id);

}
```

本例的 UserMapper 和 JPA 测试中的 UserRepository 作用完全一致，区别在于 JPA 自动生成 SQL 语句，而 MyBatis 需要开发者自己提供 SQL 语句。

④ 创建自定义服务类 Service。

程序清单：codes/06/securitymybatistest/src/main/java/org/fkit/securitymybatistest/service/UserService

```java
import java.util.ArrayList;
import java.util.List;
import org.fkit.securitymybatistest.mapper.UserMapper;
import org.fkit.securitymybatistest.pojo.FKRole;
import org.fkit.securitymybatistest.pojo.FKUser;
import org.springframework.beans.factory.annotation.Autowired;
import org.springframework.security.core.GrantedAuthority;
import org.springframework.security.core.authority.SimpleGrantedAuthority;
import org.springframework.security.core.userdetails.User;
import org.springframework.security.core.userdetails.UserDetails;
import org.springframework.security.core.userdetails.UserDetailsService;
import org.springframework.security.core.userdetails.UsernameNotFoundException;

/**
 * 需要实现UserDetailsService接口
 * 因为在Spring Security中配置相关参数需要UserDetailsService类型的数据
 * */
@Service
public class UserService implements UserDetailsService{

    @Autowired
    UserMapper userMapper;

    // 实现接口中的loadUserByUsername方法，通过该方法查询对应的用户
    @Override
    public UserDetails loadUserByUsername(String username) throws
UsernameNotFoundException {
        // 调用持久层接口findByLoginName方法查找用户，此处传入的参数实际上是loginName
        FKUser fkUser = userMapper.findByLoginName(username);
        if (fkUser == null) {
```

```
            throw new UsernameNotFoundException("用户名不存在");
        }
        // 创建 List 集合,用来保存用户权限,GrantedAuthority 对象代表赋予当前用户的权限
        List<GrantedAuthority> authorities = new ArrayList<>();
        // 获得当前用户权限集合
        List<FKRole> roles = fkUser.getRoles();
        for (FKRole role : roles) {
            // 将关联对象 Role 的 authority 属性保存为用户的认证权限
            authorities.add(new SimpleGrantedAuthority(role.getAuthority()));
        }
// 此处返回的是 org.springframework.security.core.userdetails.User 类,该类是 Spring
// Security 内部的实现
        return new User(fkUser.getUsername(), fkUser.getPassword(), authorities);
    }

}
```

本例的 UserService 和 JPA 中的 UserService 作用完全一致,区别只在于加粗的代码,注入的是 MyBatis 的持久层接口 UserMapper。

⑤ 修改 App 类。

```
@SpringBootApplication
// 扫描数据访问层接口的包名
@MapperScan("org.fkit.securitymybatistest.mapper")
public class App
{
    public static void main( String[] args )
    {
        SpringApplication.run(App.class, args);
    }
}
```

在 App 类中添加对 mapper 包扫描@MapperScan,参数是 Mapper 所在的包名,作用是让 Spring 能够扫描该包下面所有 MyBatis 的 Mapper 类。还有一种方式是直接在 Mapper 类中添加注解@Mapper。建议使用包扫描@MapperScan,否则每个 Mapper 类都要加注解,这样过于烦琐。

⑥ 测试应用。

运行 App 类的 main 方法启动项目。

在浏览器中输入 URL 测试应用。

```
http://localhost:8080
```

此时和之前的示例一样,跳转到登录页面,如图 6.1 所示。输入用户名 admin,密码 admin,单击"登录"按钮。该用户的权限是"ROLE_ADMIN"和"ROLE_DBA",跳转到 admin 页面,如图 6.8 所示。

控制台显示的执行的 SQL 语句如下:

```
    DEBUG o.f.s.mapper.UserMapper.findByLoginname  : ==> Preparing: select * from tb_user where login_name = ?
    DEBUG o.f.s.mapper.UserMapper.findByLoginName  : ==> Parameters: admin(String)
    DEBUG o.f.s.mapper.UserMapper.findRoleByUser   : ====> Preparing: SELECT id,authority FROM tb_role r,tb_user_role ur WHERE r.id = ur.role_id AND user_id = ?
    DEBUG o.f.s.mapper.UserMapper.findRoleByUser   : ====> Parameters: 1(Long)
    DEBUG o.f.s.mapper.UserMapper.findRoleByUser   : <====      Total: 2
    DEBUG o.f.s.mapper.UserMapper.findByLoginName  : <==       Total: 1
```

执行的 SQL 语句就是 Mapper 的注解中写的 SQL 语句。

读者可以根据之前示例的步骤进行测试,测试结果完全一样。

示例：基于 JDBC 的 Spring Boot Security 操作

创建一个新的 Maven 项目，命名为 securityjdbctest。按照 Maven 项目的规范，在 src/main/下新建一个名为 resources 的文件夹，并在 src/main/resources 下新建 static 和 templates 两个文件夹。该项目和 securityjpatest 的所有页面、数据库表结构、表数据完全一致，只是修改了几处代码，重点是持久层使用 JDBC 操作数据库。

① 修改 pom.xml 文件，并引入静态文件。

pom.xml 文件和之前大致相同，只是本例使用了 JDBC 操作数据库，需要引入相关模块。

```xml
<!-- 添加 spring-boot-starter-data-jdbc 模块 -->
<dependency>
    <groupId>org.springframework.boot</groupId>
    <artifactId>spring-boot-starter-jdbc</artifactId>
</dependency>
<!-- 添加 MySQL 数据库驱动 -->
<dependency>
    <groupId>mysql</groupId>
    <artifactId>mysql-connector-java</artifactId>
</dependency>
```

② 在配置文件 application.properties 中加入数据库和 JDBC 的相关配置信息。

程序清单：codes/06/securityjdbctest/src/main/resources/application.properties

```
spring.datasource.driver-class-name=com.mysql.jdbc.Driver
spring.datasource.url=jdbc:mysql://localhost:3306/springboot
spring.datasource.username=root
spring.datasource.password=
logging.level.org.springframework.security=info
spring.thymeleaf.cache=false
```

③ 创建数据访问接口。

程序清单：codes/06/securityjdbctest/src/main/java/org/fkit/securityjdbctest/repository/UserRepository

```java
import java.sql.ResultSet;
import java.sql.SQLException;
import java.util.ArrayList;
import java.util.List;
import org.fkit.securityjdbctest.pojo.FKRole;
import org.fkit.securityjdbctest.pojo.FKUser;
import org.springframework.beans.factory.annotation.Autowired;
import org.springframework.jdbc.core.JdbcTemplate;
import org.springframework.jdbc.core.RowMapper;
import org.springframework.stereotype.Repository;
import org.springframework.transaction.annotation.Transactional;

@Repository
public class UserRepository {

    // 注入 JdbcTemplate
    @Autowired
    private JdbcTemplate jdbcTemplate;

    // 根据登录名查询用户即可，不需要通过用户名和密码查询
    @Transactional(readOnly = true)
    public FKUser findByLoginName(String loginName){
        String sql = "select * from tb_user where login_name = ?";
        // 根据 loginName 查询用户
        FKUser fkUser = jdbcTemplate.queryForObject(sql,
                new Object[]{loginName}, new RowMapper<FKUser>(){
```

```java
            @Override
            public FKUser mapRow(ResultSet rs, int rowNum) throws SQLException {
                FKUser fkUser = new FKUser();
                fkUser.setId(rs.getLong("id"));
                fkUser.setLoginName(rs.getString("login_name"));
                fkUser.setPassword(rs.getString("password"));
                fkUser.setUsername(rs.getString("username"));
                return fkUser;
            }
        });
        List<FKRole> roles = new ArrayList<>();
        // 根据用户id查询用户权限
        List<Map<String,Object>> result =jdbcTemplate
          .queryForList("SELECT id,authority FROM tb_role r,tb_user_role ur "
            + "WHERE r.id = ur.role_id AND user_id = ?",new Object[]{fkUser.getId()});
        for(Map<String,Object> map : result){
          FKRole fkRole = new FKRole();
          fkRole.setId((Long)map.get("id"));
          fkRole.setAuthority((String)map.get("authority"));
          roles.add(fkRole);
        }
        // 添加用户权限
        fkUser.setRoles(roles);
        // 返回用户
        return fkUser;
    }
}
```

UserRepository 和 JPA 测试中的 UserRepository 作用完全一致，区别在于 JPA 是自动生成 SQL 语句，而 JDBC 需要开发者自己提供 SQL 语句并处理结果集，其他的工作都交给 JdbcTemplate 对象完成。

④ 创建自定义服务类 Service。

程序清单：codes/06/securityjdbctest/src/main/java/org/fkit/securityjdbctest/service/UserService

```java
import java.util.ArrayList;
import java.util.List;
import org.fkit.securityjdbctest.pojo.FKRole;
import org.fkit.securityjdbctest.pojo.FKUser;
import org.fkit.securityjdbctest.repository.UserRepository;
import org.springframework.beans.factory.annotation.Autowired;
import org.springframework.security.core.GrantedAuthority;
import org.springframework.security.core.authority.SimpleGrantedAuthority;
import org.springframework.security.core.userdetails.User;
import org.springframework.security.core.userdetails.UserDetails;
import org.springframework.security.core.userdetails.UserDetailsService;
import org.springframework.security.core.userdetails.UsernameNotFoundException;
@Service
public class UserService implements UserDetailsService{

    // 注入持久层对象UserRepository
    @Autowired
    UserRepository userRepository;

    // 实现接口中的loadUserByUsername方法，通过该方法查询对应的用户
    @Override
    public UserDetails loadUserByUsername(String username) throws
UsernameNotFoundException {
        // 调用持久层接口findByLoginName方法查找用户，此处传入的参数实际上是loginName
```

```java
//      FKUser fkUser = userRepository.findByLoginName(username);
//      System.out.println("user = " + fkUser);
        if (fkUser == null) {
            throw new UsernameNotFoundException("用户名不存在");
        }
        // 创建 List 集合，用来保存用户权限，GrantedAuthority 对象代表赋予当前用户的权限
        List<GrantedAuthority> authorities = new ArrayList<>();
        // 获得当前用户权限集合
        List<FKRole> roles = fkUser.getRoles();
        for (FKRole role : roles) {
            // 将关联对象 Role 的 authority 属性保存为用户的认证权限
            authorities.add(new SimpleGrantedAuthority(role.getAuthority()));
        }
        // 此处返回的是 org.springframework.security.core.userdetails.User 类，该类是
        // Spring Security 内部的实现
        return new User(fkUser.getUsername(), fkUser.getPassword(), authorities);
    }
}
```

UserService 和 JPA 中的 UserService 作用完全一致，区别只在于加粗的代码，注入的是 JDBC 的持久层对象 UserRepository，其中的操作是 JDBC 的实现。

读者可以根据之前示例的步骤进行测试，测试结果完全一样。

6.4 本章小结

本章主要介绍了 Spring Security 框架，以及 Spring Boot 对 Spring Security 的支持，并通过三个示例演示了在企业实际项目开发中如何使用 JPA、MyBatis 和 JDBC 进行 Spring Security 操作。

其中 JPA 和 MyBatis 的 Spring Security 操作是现代开发的主流，读者需要重点掌握。

第 7 章将使用前面所介绍的知识完成一个大型的信息管理系统。

CHAPTER 7

第 7 章
实战项目：信息管理系统

本章要点

- 信息管理系统功能改善
- 系统需求分析的基本思路
- 轻量级 Java EE 应用的分层模型
- 轻量级 Java EE 应用的总体架构及实现方案
- 根据系统需求提取系统实体
- 基于 JPA 实现持久层组件
- 实现业务逻辑层
- 基于 AOP 注解的事务
- 实现 Web 层

第7章 实战项目：信息管理系统

本章将综合运用前面章节所介绍的知识来开发一个信息管理系统。该系统包含用户管理、菜单管理、角色管理等常用的信息管理系统功能，并具有良好的维护性和扩展性，读者可以根据自己的业务需求增加相应的功能模块。

本系统采用前面介绍的 Spring Boot，持久层选择的是 JPA，该系统架构成熟，性能良好，运行稳定。Spring 的容器负责管理业务逻辑组件、持久层组件及控制层组件，充分利用 Spring 的优势，进一步增强系统的解耦，提高应用的可扩展性，降低系统重构的成本。

7.1 项目简介及系统架构

信息管理系统是办公信息自动化建设当中最常见的项目。图 7.1 显示了开发的信息管理系统界面和功能菜单。

图 7.1 "信息管理系统"界面

该项目包含了系统管理，系统管理下包含用户管理、菜单管理、角色管理等多个模块，页面使用了 jQuery 框架完成动态功能（关于 jQuery 的知识请参考疯狂软件系列图书之《疯狂前端开发讲义》）。菜单管理模块用来管理可操作的功能菜单；角色管理模块用来分配角色权限；用户管理模块包含了项目开发中常用的增删改查操作。假期管理模块是预留给读者扩展的功能模块，读者可以根据前面系统管理开发过程中所学的知识完成假期管理模块。

7.1.1 系统功能介绍

用户管理的功能包括：添加用户、查询用户（可以查询所有用户或根据条件进行模糊查询）、删除用户、修改用户、激活或冻结用户状态。

菜单管理的功能包括：添加菜单、删除菜单、修改菜单、查看下级菜单。

角色管理的功能包括：添加角色、删除角色、修改角色、绑定用户（赋予用户角色）、绑定用户操作（赋予用户可操作的功能模块）。

所有查询页面统一使用了分页处理。

7.1.2 相关技术介绍

本系统使用 Spring Boot，其中主要涉及的技术包括 Spring、Spring MVC、持久层技术 JPA、和表现层技术 jQuery、Bootstrap，考虑到现在大多数企业开发 Java 项目还在使用 JSP，所以本系统使用 JSP 作为表现层技术，本系统将这几种技术有机地结合在一起，从而构建出一个健壮的 Java EE 应用。

（1）传统表现层技术：JSP

本系统使用 JSP 作为表现层，负责收集用户请求数据以及业务数据的表示。

JSP 是最传统也最有效的表现层技术。本系统的 JSP 页面是单纯的表现层，所有的 JSP 页面不再使用 Java 脚本。结合 EL 表达式和 JSTL 标签库，JSP 可完成全部的表现层功能——数据收集、数据表示。

（2）MVC 框架

本系统使用 Spring MVC 作为 MVC 框架。Spring MVC 是一个设计优良的 MVC 框架，大有取代 Struts 2 之势。本应用的所有用户请求，包括系统的超链接和表单提交等，都不再直接发送到表现层 JSP 页面，而是必须发送给 Spring MVC 的 Controller，Spring MVC 控制所有请求的处理和转发。

通过 Controller 拦截所有请求有个好处：将所有的 JSP 页面放入 WEB-INF/路径下，可以避免用户直接访问 JSP 页面，从而提高系统的安全性。

（3）Spring 框架的作用

Spring 框架是系统的核心部分，Spring 提供的 IoC 容器是业务逻辑组件和 DAO 组件的工厂，它负责生成并管理这些实例。

借助于 Spring 的依赖注入，各组件以松耦合的方式组合在一起，组件与自己之间的依赖正是通过 Spring 的依赖注入管理的。其 Service 组件和 DAO 对象都采用面向接口编程的方式，从而降低了系统重构的成本，极好地提高了系统的可维护性、可扩展性。

应用事务采用 Spring 的注解式事务。通过注解式事务，使业务逻辑组件可以更加专注于业务的实现，从而简化开发。

（4）JPA 的作用

JPA 简化了数据库的访问，持久层只需提供接口声明，而不需要提供任何实现。大大减轻了开发者的工作量，提高了工作效率。

▶▶ 7.1.3 系统结构

本系统采用严格的 Java EE 应用结构，主要有如下几个分层。

- 表现层：由 JSP 页面组成。
- 控制层：使用 Spring MVC 技术。
- 业务层：主要由 Spring IoC 容器管理的业务逻辑组件组成。
- 持久层：由 6 个 Repository（也可以称为 DAO）组件组成。
- 领域对象层：由 6 个 Domain Object 对象组成。
- 数据库服务层：使用 MySQL 数据库存储持久化数据。

本应用中的领域对象实际上只是一些简单的 Java Bean 类，并未提供任何业务逻辑方法，所有的业务逻辑方法都由系统的业务逻辑组件来提供。这种模式简单、直接，系统分层清晰，比较适合实际项目开发。

▶▶ 7.1.4 系统的功能模块

本系统重点是系统管理模块，该模块可以大致分为三个小模块：用户管理、菜单管理、角色管理，其主要业务逻辑通过 IdentityService 业务逻辑组件实现，因此可以使用这个业务逻辑组件来封装 Repository 组件。

> **注意**
> 通常建议按细粒度的模块来设计 Service 组件，让业务逻辑组件作为 Repository 组件的门面，这符合门面模式的设计。同时让 Repository 组件负责系统持久化逻辑，可以将系统在持久化技术这个维度上的变化独立出去，而业务逻辑组件负责业务逻辑这个维度的改变。

系统以业务逻辑组件作为 Repository 组件的门面，封装这些 Repository 组件，业务逻辑组件底层依赖于这些 Repository 组件，向上实现系统的业务逻辑功能。

本系统主要有如下 6 个 Repository 对象。
- UserRepository：提供对 oa_id_user 表的基本操作。
- DeptRepository：提供对 oa_id_dept 表的基本操作。
- JobRepository：提供对 oa_id_job 表的基本操作。
- ModuleRepository：提供对 oa_id_module 表的基本操作。
- RoleRepository：提供对 oa_id_role 表的基本操作。
- PopedomRepository：提供对 oa_id_popedom 表的基本操作。

本系统还提供一个业务逻辑组件。
- IdentityService：提供所有的业务逻辑功能的实现。

7.2 配置文件

创建一个新的 Maven Web 项目，命名为 oa，按照 Maven 项目的规范，在 src/main/下新建一个名为 resources 的文件夹，并在 src/main/resources 下再新建 static 和 templates 两个文件夹。将项目所需的静态文件放到 static 目录下。

程序清单：codes/07/oa/pom.xml

```xml
<project xmlns="http://maven.apache.org/POM/4.0.0" xmlns:xsi="http://www.w3.org/2001/XMLSchema-instance"
    xsi:schemaLocation="http://maven.apache.org/POM/4.0.0 http://maven.apache.org/xsd/maven-4.0.0.xsd">
    <modelVersion>4.0.0</modelVersion>

    <groupId>org.fkit</groupId>
    <artifactId>oa</artifactId>
    <packaging>war</packaging>
    <version>0.0.1-SNAPSHOT</version>

    <name>oa Maven Webapp</name>
    <url>http://maven.apache.org</url>

    <!--
        spring-boot-starter-parent 是 Spring Boot 的核心启动器，
        包含了自动配置、日志和 YAML 等大量默认的配置，大大简化了开发工作。
        之后引入相关的依赖模块就不需要添加版本号配置，
        Spring Boot 会自动选择最合适的版本进行添加。
    -->
    <parent>
        <groupId>org.springframework.boot</groupId>
        <artifactId>spring-boot-starter-parent</artifactId>
        <version>2.0.0.RELEASE</version>
```

```xml
        </parent>

        <properties>
            <project.build.sourceEncoding>UTF-8</project.build.sourceEncoding>
<project.reporting.outputEncoding>UTF-8</project.reporting.outputEncoding>
            <java.version>1.8</java.version>
            <json.version>2.4</json.version>
        </properties>

        <dependencies>
            <!-- 添加 spring-boot-starter-web 依赖 -->
            <dependency>
                <groupId>org.springframework.boot</groupId>
                <artifactId>spring-boot-starter-web</artifactId>
            </dependency>

            <!-- 添加 MySQL 数据库依赖 -->
            <dependency>
                <groupId>mysql</groupId>
                <artifactId>mysql-connector-java</artifactId>
            </dependency>

            <!-- 添加 spring-boot-starter-data-jpa 依赖 -->
            <dependency>
                <groupId>org.springframework.boot</groupId>
                <artifactId>spring-boot-starter-data-jpa</artifactId>
            </dependency>

        <!-- 添加 servlet 依赖 -->
        <dependency>
            <groupId>javax.servlet</groupId>
              <artifactId>javax.servlet-api</artifactId>
            <scope>provided</scope>
        </dependency>

            <!-- 添加 JSTL 依赖 -->
            <dependency>
                <groupId>javax.servlet</groupId>
                <artifactId>jstl</artifactId>
            </dependency>

            <!-- 添加 spring-boot-starter-tomcat 依赖 -->
            <dependency>
                <groupId>org.springframework.boot</groupId>
                <artifactId>spring-boot-starter-tomcat</artifactId>
                <scope>provided</scope>
            </dependency>

            <!-- 添加使用 jsp 必需的 tomcat-embed-jasper 依赖 -->
            <dependency>
                <groupId>org.apache.tomcat.embed</groupId>
                <artifactId>tomcat-embed-jasper</artifactId>
                <scope>provided</scope>
            </dependency>

        <dependency>
          <groupId>junit</groupId>
          <artifactId>junit</artifactId>
          <scope>test</scope>
        </dependency>

        </dependencies>

        <build>
```

```xml
        <finalName>oa</finalName>
    </build>

</project>
```

程序清单：codes/07/oa/src/main/resources/application.properties

```properties
###########################################################
### 数据源信息配置
###########################################################
# 数据库地址
spring.datasource.url = jdbc:mysql://localhost:3306/oa_db
# 用户名
spring.datasource.username = root
# 密码
spring.datasource.password =
# 数据库驱动
spring.datasource.driverClassName = com.mysql.jdbc.Driver
# 指定连接池中最大的活跃连接数
spring.datasource.max-active=20
# 指定连接池最大的空闲连接数
spring.datasource.max-idle=8
# 指定必须保持连接的最小值
spring.datasource.min-idle=8
# 指定启动连接池时，初始建立的连接数量
spring.datasource.initial-size=10
###########################################################
### JPA 持久化配置
###########################################################
# 指定数据库的类型
spring.jpa.database = MySQL
# 指定是否需要在日志中显示 SQL 语句
spring.jpa.show-sql = true
# 指定自动创建|更新|验证数据库表结构等配置，配置成 update
# 表示如果数据库中存在持久化类对应的表就不创建，不存在就创建对应的表
spring.jpa.hibernate.ddl-auto = update
# 指定命名策略
spring.jpa.hibernate.naming-strategy = org.hibernate.cfg.ImprovedNamingStrategy
# 指定数据库方言
spring.jpa.properties.hibernate.dialect = org.hibernate.dialect.MySQL5Dialect

spring.mvc.view.prefix: /WEB-INF/jsp/
spring.mvc.view.suffix: .jsp
```

注意

> JPA 只会帮助开发者创建表，而不会创建数据库。读者需要先在 MySQL 当中手动创建数据库。

7.3 持久化类

通过使用 JPA 持久层，可以避免使用传统的 JDBC 方式来操作数据库，并且可以直接面向对象操作数据。

▶▶ 7.3.1 设计持久化实体

面向对象分析，是指根据系统需求提取应用中的对象，将这些对象抽象成类，再抽取出需

要持久化保存的类，这些需要持久化保存的类就是持久化对象（PO）。本系统设计了 6 个持久化类。

- User：对应用户，包括用户 ID、姓名、密码、性别、所属部门、工作岗位、电话号码、邮箱、QQ 号码、状态等属性。
- Dept：对应部门，包括部门 ID、部门名称及备注等属性。
- Job：对应职位，包括职位编号、职位名称及备注等属性。
- Module：对应模块，包括模块 ID、模块名称、操作链接、备注、创建日期及修改日期等属性。
- Role：对应角色，包括角色 ID、备注、发布日期等属性。
- Popedom：对应权限，包括权限 ID、权限对应的模块、权限对应的操作、权限对应的角色及创建时间等属性。

在富领域模式的设计中，这 6 个 PO 对象也应该包含系统的业务逻辑方法，即每一个 PO 对应一个 Service，也就是使用领域模型对象来为它们建模；但本应用不打算为它们提供任何业务逻辑方法，而是将所有的业务逻辑方法放到业务逻辑组件中实现。这样，系统中的领域对象就会十分简捷，它们都是单纯的数据类，不需要考虑到底应该包含哪些业务逻辑方法，因此开发起来非常便捷；而系统的所有业务逻辑都由业务逻辑组件负责实现，可以将业务逻辑的变化限制在业务逻辑层内，从而避免扩散到两个层，因此降低了系统的开发难度。

客观世界中的对象不是孤立存在的，以上 6 个 PO 类也不是孤立存在的，它们之间存在复杂的关联关系，分析关联关系是面向对象分析的必要步骤。

这 6 个 PO 的关系如下：

Dept 和 User 之间存在 1—N 的关系，即一个 Dept（部门）可以有多个 User（用户），它们的关系通过主外键关联。

User 和 Dept 之间存在 N—1 的关系，即一个 User（用户）只属于一个 Dept（部门），它们的关系通过主外键关联。

User 和 Job 之间存在 N—1 的关系，即一个 User（用户）只能有一个 Job（职位），它们的关系通过主外键关联。

Module 和 Role 是 N—N 的关系，即一个 Module（模块）可以属于多个 Role（角色），一个 Role（角色）中也可以有多个 Module（模块），它们的关系通过对象 Popedom 关联。

User 和 Role 是 N—N 的关系，即一个 User（用户）可以分配多个 Role（角色），一个 Role（角色）可以分配给多个 User（用户），它们的关系通过中间表 oa_id_user_role 关联。

7.3.2 创建持久化实体类

使用 JPA 持久化框架，持久化对象之间的关联关系都是以成员变量的方式表现出来，当然，这些成员变量同样需要 setter 和 getter 方法的支持，持久化类之间的关联关系通常对应数据库里的主、外键进行约束。

除此之外，持久化对象还有自己的普通类型的成员变量，这些成员变量通常对应数据库的字段。

下面是 6 个持久化类的源代码。

程序清单：codes/07/oa/src/main/java/org/fkit/oa/identity/domain/Dept

```java
import java.io.Serializable;
import java.util.Date;
import javax.persistence.*;
```

```java
@Entity @Table(name="OA_ID_DEPT")
public class Dept implements Serializable {

    private static final long serialVersionUID = 678100638005497362L;
    /** ID     NUMBER      编号     PK 主键自增长*/
    @Id @GeneratedValue(strategy=GenerationType.AUTO)
    @Column(name="ID")
    private Long id;

    /** NAME     VARCHAR2(50) 部门名称*/
    @Column(name="NAME", length=50)
    private String name;

    /** REMARK     VARCHAR2(500)     备注     */
    @Column(name="REMARK", length=500)
    private String remark;

    /** MODIFIER VARCHAR2(50) 修改人     FK(OA_ID_USER) N-1   */
    @ManyToOne(fetch=FetchType.LAZY, targetEntity=User.class)
    @JoinColumn(name="MODIFIER", referencedColumnName="USER_ID",
            foreignKey=@ForeignKey(name="FK_DEPT_MODIFIER")) // 更改外键约束名
    private User modifier;

    /** MODIFY_DATE   DATE    修改时间*/
    @Column(name="MODIFY_DATE")
    @Temporal(TemporalType.TIMESTAMP)
    private Date modifyDate;

    /** CREATER     VARCHAR2(50) 创建人     FK(OA_ID_USER)*/
    @ManyToOne(fetch=FetchType.LAZY, targetEntity=User.class)
    @JoinColumn(name="CREATER", referencedColumnName="USER_ID",
            foreignKey=@ForeignKey(name="FK_DEPT_CREATER")) // 更改外键约束名
    private User creater;

    /** CREATE_DATE    DATE    创建时间*/
    @Column(name="CREATE_DATE")
    @Temporal(TemporalType.TIMESTAMP)
    private Date createDate;

    // 省略构造器、set 和 get 方法
}
```

程序清单：codes/07/oa/src/main/java/org/fkit/oa/identity/domain/User

```java
import java.io.Serializable;
import java.util.Date;
import java.util.HashSet;
import java.util.Set;
import javax.persistence.*;

@Entity @Table(name="OA_ID_USER",
        indexes={@Index(columnList="NAME", name="IDX_USER_NAME")})
//@Cache(usage=CacheConcurrencyStrategy.READ_WRITE)
public class User implements Serializable {

    /** 用户 ID     PK，大小写英文和数字 */
    @Id @Column(name="USER_ID", length=50)
    private String userId;
    /** 密码     MD5 加密 */
    @Column(name="PASS_WORD", length=50)
    private String passWord;
    /** 姓名 */
```

```java
@Column(name="NAME", length=50)
private String name;
/** 性别    1:男 2:女 */
@Column(name="SEX")
private Short sex = 1;
/** 用户与部门存在多对一关联    部门    FK(OA_ID_DEPT) */
@ManyToOne(fetch=FetchType.LAZY, targetEntity=Dept.class)
@JoinColumn(name="DEPT_ID", referencedColumnName="ID",
        foreignKey=@ForeignKey(name="FK_USER_DEPT"))
private Dept dept;   // select u from User u where u.dept.id = ?

/** 用户与职位存在多对一关联    职位    FK(OA_ID_JOB) */
@ManyToOne(fetch=FetchType.LAZY, targetEntity=Job.class)
@JoinColumn(name="JOB_CODE", referencedColumnName="CODE",
        foreignKey=@ForeignKey(name="FK_USER_JOB"))
private Job job;
/** 邮箱 */
@Column(name="EMAIL", length=50)
private String email;
/** 电话号码 */
@Column(name="TEL", length=50)
private String tel;
/** 手机号码 */
@Column(name="PHONE", length=50)
private String phone;
/** QQ 号码 */
@Column(name="QQ_NUM", length=50)
private String qqNum;
/** 问题编号 */
@Column(name="QUESTION")
private Short question;
/** 回答结果 */
@Column(name="ANSWER", length=200)
private String answer;
/** 状态    0 新建，1 审核，2 不通过审核，3 冻结    */
@Column(name="STATUS")
private Short status = 0;
/** 用户创建人与用户存在多对一关联(FK(OA_ID_USER)) */
@ManyToOne(fetch=FetchType.LAZY, targetEntity=User.class)
@JoinColumn(name="CREATER", referencedColumnName="USER_ID",
        foreignKey=@ForeignKey(name="FK_USER_CREATER"))  // 更改外键约束名
private User creater;
/** 创建时间 */
@Column(name="CREATE_DATE")
@Temporal(TemporalType.TIMESTAMP)
private Date createDate;
/** 用户修改人与用户存在多对一关联(FK(OA_ID_USER)) */
@ManyToOne(fetch=FetchType.LAZY, targetEntity=User.class)
@JoinColumn(name="MODIFIER", referencedColumnName="USER_ID",
        foreignKey=@ForeignKey(name="FK_USER_MODIFIER"))  // 更改外键约束名
private User modifier;
/** 修改时间 */
@Column(name="MODIFY_DATE")
@Temporal(TemporalType.TIMESTAMP)
private Date modifyDate;
/** 部门审核人与用户存在多对一关联(FK(OA_ID_USER)) */
@ManyToOne(fetch=FetchType.LAZY, targetEntity=User.class)
@JoinColumn(name="CHECKER", referencedColumnName="USER_ID",
        foreignKey=@ForeignKey(name="FK_USER_CHECKER"))  // 更改外键约束名
private User checker;
/** 审核时间 */
@Column(name="CHECK_DATE")
```

```java
    @Temporal(TemporalType.TIMESTAMP)
    private Date checkDate;

    /** 用户与角色存在多对多关联 */
    @ManyToMany(fetch=FetchType.LAZY, targetEntity=Role.class, mappedBy="users")
    private Set<Role> roles = new HashSet<>();
    // 省略构造器、set 和 get 方法
}
```

程序清单：codes/07/oa/src/main/java/org/fkit/oa/identity/domain/Job

```java
import java.io.Serializable;

import javax.persistence.*;

import org.hibernate.annotations.Cache;
import org.hibernate.annotations.CacheConcurrencyStrategy;

@Entity @Table(name="OA_ID_JOB")
//@Cache(usage=CacheConcurrencyStrategy.READ_WRITE)
public class Job implements Serializable {

    private static final long serialVersionUID = 4594973777750274376L;
    /**
     * CODE      VARCHAR2(100)  代码 PK 主键
     * (0001...0002) 四位为模块；
     * (00010001..) 八位为操作
     */
    @Id @Column(name="CODE", length=100)
    private String code;
    /** NAME VARCHAR2(50)  名称 */
    @Column(name="NAME", length=50)
    private String name;
    /** REMARK    VARCHAR2(300)  职位说明*/
    @Column(name="REMARK", length=300)
    private String remark;
    // 省略构造器、set 和 get 方法
}
```

程序清单：codes/07/oa/src/main/java/org/fkit/oa/identity/domain/Module

```java
import java.io.Serializable;
import java.util.Date;

import javax.persistence.*;

@Entity @Table(name="OA_ID_MODULE")
public class Module implements Serializable {

    private static final long serialVersionUID = 5139796285142133024L;
    /**
     * CODE     VARCHAR2(100)       代码     PK 主键
     * (0001...0002) 四位为模块；
     * (00010001..) 八位为操作
     */
    @Id @Column(name="CODE", length=100)
    private String code;
    /** 名称 */
    @Column(name="NAME", length=50)
    private String name;
    /** 操作链接 */
    @Column(name="URL", length=100)
    private String url;
```

```java
/** 备注    */
@Column(name="REMARK", length=500)
private String remark;
/** 模块修改人与用户存在多对一关联(FK(OA_ID_USER)) */
@ManyToOne(fetch=FetchType.LAZY, targetEntity=User.class)
@JoinColumn(name="MODIFIER", referencedColumnName="USER_ID",
        foreignKey=@ForeignKey(name="FK_MODULE_MODIFIER")) // 更改外键约束名
private User modifier;
/** 修改时间 */
@Column(name="MODIFY_DATE")
@Temporal(TemporalType.TIMESTAMP)
private Date modifyDate;
/** 模块创建人与用户存在多对一关联(FK(OA_ID_USER)) */
@ManyToOne(fetch=FetchType.LAZY, targetEntity=User.class)
@JoinColumn(name="CREATER", referencedColumnName="USER_ID",
        foreignKey=@ForeignKey(name="FK_MODULE_CREATER")) // 更改外键约束名
private User creater;
/** 创建时间 */
@Column(name="CREATE_DATE")
@Temporal(TemporalType.TIMESTAMP)
private Date createDate;
// 省略构造器、set 和 get 方法
}
```

程序清单：codes/07/oa/src/main/java/org/fkit/oa/identity/domain/Role

```java
import java.io.Serializable;
import java.util.Date;
import java.util.HashSet;
import java.util.Set;

import javax.persistence.*;

import org.hibernate.annotations.Cache;
import org.hibernate.annotations.CacheConcurrencyStrategy;

@Entity @Table(name="OA_ID_ROLE")
//@Cache(usage=CacheConcurrencyStrategy.READ_WRITE)
public class Role implements Serializable {

    private static final long serialVersionUID = 6837526111700641932L;
    /** 编号    PK 主键自增长 */
    @Id @GeneratedValue(strategy=GenerationType.AUTO)
    @Column(name="ID")
    private Long id;
    /** 角色名字 */
    @Column(name="NAME", length=50)
    private String name;
    /** 备注    */
    @Column(name="REMARK", length=500)
    private String remark;
    /** 角色创建人与用户存在多对一关联(FK(OA_ID_USER)) */
    @ManyToOne(fetch=FetchType.LAZY, targetEntity=User.class)
    @JoinColumn(name="CREATER", referencedColumnName="USER_ID",
            foreignKey=@ForeignKey(name="FK_ROLE_CREATER")) // 更改外键约束名
    private User creater;
    /** 创建时间 */
    @Column(name="CREATE_DATE")
    @Temporal(TemporalType.TIMESTAMP)
    private Date createDate;
    /** 角色修改人与用户存在多对一关联(FK(OA_ID_USER)) */
    @ManyToOne(fetch=FetchType.LAZY, targetEntity=User.class)
    @JoinColumn(name="MODIFIER", referencedColumnName="USER_ID",
```

```java
                foreignKey=@ForeignKey(name="FK_ROLE_MODIFIER"))  // 更改外键约束名
    private User modifier;
    /** 修改时间 */
    @Column(name="MODIFY_DATE")
    @Temporal(TemporalType.TIMESTAMP)
    private Date modifyDate;

    /** 角色与用户存在多对多关联 */
    @ManyToMany(fetch=FetchType.LAZY, targetEntity=User.class)
    @JoinTable(name="OA_ID_USER_ROLE", joinColumns=@JoinColumn(name="ROLE_ID",
        referencedColumnName="ID"),
                 inverseJoinColumns=@JoinColumn(name="USER_ID",
        referencedColumnName="USER_ID"))
    private Set<User> users = new HashSet<>();
    // 省略构造器、set 和 get 方法
}
```

程序清单：codes/07/oa/src/main/java/org/fkit/oa/identity/domain/Popedom

```java
import java.io.Serializable;
import java.util.Date;

import javax.persistence.*;

@Entity @Table(name="OA_ID_POPEDOM")
public class Popedom implements Serializable {
    // Popedom p = new Popedom(); // 一行数据
    private static final long serialVersionUID = -12461070000138494011L;
    /** 编号    PK 主键自增长 */
    @Id @GeneratedValue(strategy=GenerationType.AUTO)
    @Column(name="ID")
    private Long id;
    /** 权限与模块存在多对一关联  模块代码 FK(OA_ID_MODULE) */
    @ManyToOne(fetch=FetchType.LAZY, targetEntity=Module.class)
    @JoinColumn(name="MODULE_CODE", referencedColumnName="CODE",
         foreignKey=@ForeignKey(name="FK_POPEDOM_MODULE"))  // 更改外键约束名
    private Module module;
    /** 权限与操作存在多对一关联  操作代码 FK(OA_ID_MODULE) */
    @ManyToOne(fetch=FetchType.LAZY, targetEntity=Module.class)
    @JoinColumn(name="OPERA_CODE", referencedColumnName="CODE",
         foreignKey=@ForeignKey(name="FK_POPEDOM_OPERA"))  // 更改外键约束名
    private Module opera;
    /** 权限与角色存在多对一关联  角色  FK(OA_ID_ROLE) */
    @ManyToOne(fetch=FetchType.LAZY, targetEntity=Role.class)
    @JoinColumn(name="ROLE_ID", referencedColumnName="ID",
         foreignKey=@ForeignKey(name="FK_POPEDOM_ROLE"))  // 更改外键约束名
    private Role role;
    /** 权限创建人与用户存在多对一关联(FK(OA_ID_USER)) */
    @ManyToOne(fetch=FetchType.LAZY, targetEntity=User.class)
    @JoinColumn(name="CREATER", referencedColumnName="USER_ID",
         foreignKey=@ForeignKey(name="FK_POPEDOM_CREATER"))  // 更改外键约束名
    private User creater;
    /** 创建时间 */
    @Column(name="CREATE_DATE")
    @Temporal(TemporalType.TIMESTAMP)
    private Date createDate;
    // 省略构造器、set 和 get 方法
}
```

7.3.3 导入初始数据

在 MySQL 数据库系统中创建一个名为"oa_db"的数据库。

像之前章节的项目那样修改 App 类。

运行 App 类，运行成功之后，"oa_db" 数据库会生成 7 张表。

- oa_id_dept：部门表，映射 Dept 类。
- oa_id_user：用户表，映射 User 类。
- oa_id_job：职位表，映射 Job 类。
- oa_id_module：模块表，映射 Module 类。
- oa_id_role：角色表，映射 Role 类。
- oa_id_popedom：权限表，映射 Popedom 类。
- oa_id_user_role：用户和角色的中间表。

导入配套资源文件中的"oa_测试数据脚本.sql"，向 7 张表中插入测试数据，用于项目功能测试。具体的 SQL 语句请参考"oa_测试数据脚本.sql"。

7.4 定义 Repository 接口实现 Repository 持久层

Spring Data 建议定义接口完成 SQL 语句的操作，该接口可以直接作为 Repository 组件（又称为 DAO 组件）使用。当使用 Repository 模式时，既能体现业务逻辑组件封装 Repository 组件的门面模式，也可分离业务逻辑组件和 Repository 组件的功能：业务逻辑组件负责业务逻辑的变化，而 Repository 组件负责持久化技术的变化，这正是桥接模式的应用。

引入 Repository 模式后，每个 Repository 组件包含了数据库的访问逻辑；每个 Repository 组件可对一个数据库表完成基本的 CRUD 等操作。

在 Spring Data 当中，基本的增删改查已经在父类 JpaRepository 中完成，自定义接口中只需通过@Query 注解完成个性化查询即可。通过 Repository 接口完成数据库的操作，这种简单的实现较之传统的 JDBC 持久化访问，简直不可同日而语。

注意

在学习框架的过程中也许会有少许的坎坷，不过一旦掌握了框架的作用，将大幅提高应用的开发效率，而且好的框架所倡导的软件架构还会提高开发者的架构设计知识。

下面是 6 个 Repository 接口的源代码。

程序清单：codes/07/oa/src/main/java/org/fkit/oa/identity/repository/DeptRepository

```java
import java.util.Map;
import org.fkit.oa.identity.domain.Dept;
import org.springframework.data.jpa.repository.JpaRepository;
import org.springframework.data.jpa.repository.Query;

public interface DeptRepository extends JpaRepository<Dept, Long>{

    @Query("select new Map(p.id as code , p.name as name) from Dept p")
    public List<Map<String, Object>> findDepts();

}
```

DeptRepository 接口继承 JpaRepository<Dept, Long>接口，定义了一个查询所有部门的方法。

程序清单：codes/07/oa/src/main/java/org/fkit/oa/identity/repository/UserRepository

```java
import java.util.List;
import java.util.List;
import org.fkit.oa.identity.domain.User;
import org.springframework.data.jpa.repository.JpaRepository;
import org.springframework.data.jpa.repository.JpaSpecificationExecutor;
import org.springframework.data.jpa.repository.Query;

public interface UserRepository extends JpaRepository<User, String> ,
JpaSpecificationExecutor<User>{

    @Query("select u.userId from User u where u.userId not in(select u.userId from User u inner join u.roles r where r.id = ?1)")
    List<String> getRolesUsers(Long id);

    @Query("select u.userId from User u inner join u.roles r where r.id = ?1")
    List<String> findRoleUsers(Long id);
}
```

UserRepository 接口继承 JpaRepository<User, String>接口和 JpaSpecificationExecutor<User>接口，定义了两个查询用户角色的方法。

程序清单：codes/07/oa/src/main/java/org/fkit/oa/identity/repository/JobRepository

```java
import java.util.Map;
import org.fkit.oa.identity.domain.Job;
import org.springframework.data.jpa.repository.JpaRepository;
import org.springframework.data.jpa.repository.Query;

public interface JobRepository extends JpaRepository<Job, String>{
    @Query("select new Map(j.code as code ,j.name as name) from Job j")
    public List<Map<String, Object>> findJobs() throws Exception ;
}
```

JobRepository 接口继承 JpaRepository<Job, String>接口，定义了查询所有部门的方法。

程序清单：codes/07/oa/src/main/java/org/fkit/oa/identity/repository/ModuleRepository

```java
import java.util.List;
import java.util.List;
import org.fkit.oa.identity.domain.Module;
import org.springframework.data.jpa.repository.JpaRepository;
import org.springframework.data.jpa.repository.JpaSpecificationExecutor;
import org.springframework.data.jpa.repository.Modifying;
import org.springframework.data.jpa.repository.Query;
import org.springframework.data.repository.query.Param;

public interface ModuleRepository extends JpaRepository<Module, String> ,
JpaSpecificationExecutor<Module>{

    @Modifying
    @Query("delete Module m where m.code like ?1")
    public void setCode(String code) ;

    @Query("select m from Module m where m.code like :parentCode and length(m.code) = :sonCodeLen")
    public List<Module> findModules(@Param("parentCode")String parentCode, @Param("sonCodeLen")int sonCodeLen);

    @Query("select Max(m.code) from Module m where m.code like :parentCode and
```

```
length(m.code) = :sonCodeLen ")
    public String findUniqueEntity(@Param("parentCode")String parentCode, @Param
("sonCodeLen")int sonCodeLen);
    }
```

ModuleRepository 接口继承 JpaRepository< Module, String>接口，定义了查询所有模块的方法。

程序清单：codes/07/oa/src/main/java/org/fkit/oa/identity/repository/RoleRepository

```
import org.fkit.oa.identity.domain.Role;
import org.springframework.data.jpa.repository.JpaRepository;
import org.springframework.data.jpa.repository.JpaSpecificationExecutor;

public interface RoleRepository extends JpaRepository<Role, Long> ,
JpaSpecificationExecutor<Role>{

}
```

RoleRepository 接口继承 JpaRepository<Role, Long >接口和 JpaSpecificationExecutor<Role>。

程序清单：codes/07/oa/src/main/java/org/fkit/oa/identity/repository/PopedomRepository

```
import java.util.List;
import org.fkit.oa.identity.domain.Popedom;
import org.springframework.data.jpa.repository.JpaRepository;
import org.springframework.data.jpa.repository.Modifying;
import org.springframework.data.jpa.repository.Query;
import org.springframework.data.repository.query.Param;

public interface PopedomRepository extends JpaRepository<Popedom, Long>{

    @Query("select p.opera.code from Popedom p where p.role.id = :id and p.module.
code = :parentCode")
    public List<String> findByIdAndParentCode(@Param("id")long id, @Param
("parentCode")String parentCode);

    @Modifying
    @Query("delete Popedom p where p.role.id = :id and p.module.code = :parentCode")
    public void setByIdAndParentCode(@Param("id") long id, @Param("parentCode")
String parentCode);

    @Query("select distinct p.module.code from Popedom p where "
        + "p.role.id in(select r.id from Role r "
        + "inner join r.users u where u.userId = ?1 ) "
        + "order by p.module.code asc")
    public List<String> getUserPopedomModuleCodes(String userId);

    @Query("select distinct p.opera.code from Popedom p "
        + "where p.role.id in(select r.id from Role r "
        + "inner join r.users u where u.userId = ?1 ) order by p.opera.code asc")
    public List<String> getUserPopedomOperasCodes(String userId);

}
```

PopedomRepository 接口继承 JpaRepository<Popedom, Long >接口，定义了权限的相关查询方法。

7.5 实现 Service 持久层

本系统只使用了一个业务逻辑组件：IdentityService。该组件作为门面封装 6 个 Repository 组件，系统使用这个业务逻辑组件将这些 Repository 对象封装在一起。

7.5.1 业务逻辑组件的设计

业务逻辑是 Repository 组件的门面，所以也可以理解为业务逻辑组件需要依赖于 Repository 组件。

在 IdentityService 接口里定义了大量的业务方法，这些方法的实现依赖于 Repository 组件。由于每个业务方法要涉及多个 Repository 操作，其 Repository 操作是单条数据记录的操作，而业务逻辑方法的访问，则需要涉及多个 Repository 操作，因此每个业务逻辑方法可能需要涉及多条记录的访问。

业务逻辑组件面向 Repository 接口编程，可以让业务逻辑组件从 Repository 组件的实现中分离。因此业务逻辑组件只关心业务逻辑的实现，无须关心数据访问逻辑的实现。

7.5.2 实现业务逻辑组件

业务逻辑组件负责实现系统所需的业务方法，系统有多少个业务方法，业务逻辑组件就提供多少个对应方法，业务逻辑方法完全由业务逻辑组件负责实现。

业务逻辑组件只负责业务逻辑上的变化，而持久层上的变化则交给 DAO 层负责，因此业务逻辑组件必须依赖于 DAO 组件。

为了简化分页功能，设计了一个分页的 JSP 标签，只需在页面使用分页标签，就可以完成所有页面的分页功能。

程序清单：codes/07/oa/src/main/java/org/fkit/oa/common/util/pager/PageModel

```java
// 分页实体
public class PageModel {

    /** 分页中默认一页4条数据 */
    public static final int PAGE_DEFAULT_SIZE = 4;

    /** 分页总数据条数 */
    private long recordCount;
    /** 当前页面 */
    private int pageIndex ;
    /** 每页分多少条数据 */
    private int pageSize = PAGE_DEFAULT_SIZE;

    /** 总页数 */
    private int totalSize;

    public long getRecordCount() {
        this.recordCount = this.recordCount <= 0 ? 0:this.recordCount;
        return recordCount;
    }
    public void setRecordCount(long recordCount) {
        this.recordCount = recordCount;
    }
    public int getPageIndex() {
//        this.pageIndex = this.pageIndex>=this.getTotalSize()?this.getTotalSize():this.pageIndex;
        this.pageIndex = this.pageIndex <= 0?1:this.pageIndex;
//        /** 判断当前页面是否超过了总页数:如果超过，默认将最后一页作为当前页 */
        return pageIndex;
    }
    public void setPageIndex(int pageIndex) {
        this.pageIndex = pageIndex;
    }
```

```java
    public int getPageSize() {
        this.pageSize = this.pageSize <= PAGE_DEFAULT_SIZE?PAGE_DEFAULT_SIZE:this.pageSize;
        return pageSize;
    }
    public void setPageSize(int pageSize) {
        this.pageSize = pageSize;
    }

    public int getTotalSize() {
        if(this.getRecordCount() <=0){
            totalSize = 0 ;
        }else{
            totalSize = (int) ((this.getRecordCount() -1)/this.getPageSize() + 1);
        }
        return totalSize;
    }

    public int getFirstLimitParam(){
        return (this.getPageIndex()-1)*this.getPageSize() ;
    }
}
```

程序清单：codes/07/oa/src/main/java/org/fkit/oa/common/util/pager/PageModel

```java
import java.io.IOException;
import javax.servlet.jsp.JspException;
import javax.servlet.jsp.tagext.SimpleTagSupport;

public class PagerTag extends SimpleTagSupport {

    /** 定义请求URL中的占位符常量 */
    private static final String TAG = "{0}";

    /** 当前页码 */
    private int pageIndex;
    /** 每页显示的数量 */
    private int pageSize;
    /** 总记录条数 */
    private int recordCount;
    /** 请求URL page.action?pageIndex={0}*/
    private String submitUrl;
    /** 样式 */
    private String style = "sabrosus";

    /** 定义总页数 */
    private int totalPage = 0;

    /**  在页面上引用自定义标签就会触发一个标签处理类   */
    @Override
    public void doTag() throws JspException, IOException {
        System.out.println("========================");

        System.out.println("-"+pageIndex+"-"+pageSize);
        /** 定义StringBuilder用来拼接字符串 */
        StringBuilder res = new StringBuilder();

        res.append("<center>\n" +
            "\t\t<p style=\"text-align: center;\">\n" +
            "\t\t\t<!-- 设计导航-->\n" +
            "\t\t\t<nav class=\"nav form-inline\">\n" +
            "\t\t\t\t <ul class=\"pagination alin\">");
```

```java
/** 定义它拼接中间的页码 */
StringBuilder str = new StringBuilder();
/** 判断总记录条数 */
if (recordCount > 0){
    /** 需要显示分页标签,计算出总页数、需要分多少页
     *  1500 条数据,每页 15 条, 1500/15 = 100 页
     * */
    totalPage = (this.recordCount - 1) / this.pageSize + 1;

    /** 判断上一页或下一页需不需要加 a 标签 */
    if (this.pageIndex == 1){ // 首页
        str.append("<li class=\"disabled\" ><a href=\"#\">上一页</a></li>");

        /** 计算中间的页码 */
        this.calcPage(str);

        /** 下一页需不需要 a 标签 */
        if (this.pageIndex == totalPage){
            /** 只有一页 */
            str.append("<li class=\"disabled\" ><a href=\"#\">下一页</a></li>");
        }else{
            String tempUrl = this.submitUrl.replace(TAG, String.valueOf(pageIndex + 1));
            str.append("<li><a href='"+tempUrl+"'>下一页</a></li>");
        }
    }else if (this.pageIndex == totalPage){ // 尾页
        String tempUrl = this.submitUrl.replace(TAG, String.valueOf(pageIndex - 1));
        str.append("<li><a href='"+tempUrl+"'>上一页</a></li>");

        /** 计算中间的页码 */
        this.calcPage(str);

        str.append("<li class=\"disabled\" ><a href=\"#\">下一页</a></li>");
    }else{ // 中间
        String tempUrl = this.submitUrl.replace(TAG, String.valueOf(pageIndex - 1));
        str.append("<li><a href='"+tempUrl+"'>上一页</a></li>");

        /** 计算中间的页码 */
        this.calcPage(str);

        tempUrl = this.submitUrl.replace(TAG, String.valueOf(pageIndex + 1));
        str.append("<li><a href='"+tempUrl+"'>下一页</a></li>");
    }

    res.append(str);

    /** 开始条数 */
    int startNum = (this.pageIndex - 1) * this.pageSize + 1;
    /** 结束条数 */
    int endNum = (this.pageIndex == this.totalPage) ? this.recordCount : this.pageIndex * this.pageSize;

    res.append("<li><a style=\"background-color:#D4D4D4;\" href=\"#\">共<font color='red'>"+this.recordCount+"</font>条记录,当前显示"+startNum+"-"+endNum+"条记录</a> </li>");

    res.append("<div class=\"input-group\">\n" +
            "\t\t\t\t\t\t\t\t\t<input id='pager_jump_page_size' value='"+this.pageIndex+"' type=\"text\" style=\"width: 60px;text-align: center;\"
```

```
class=\"form-control\" placeholder=\""+this.pageIndex+"\"\">\n" +
                            "\t\t\t\t\t\t\t\t\t   <span class=\"input-group-btn\">\n" +
                            "\t\t\t\t\t\t\t\t\t     <button class=\"btn btn-info\"
id='pager_jump_btn' type=\"button\">GO</button>\n" +
                            "\t\t\t\t\t\t\t   </span>\n" +
                            "\t\t\t\t\t  \t\t\t\t </div>");

            res.append("<script type='text/javascript'>");
            res.append("  document.getElementById('pager_jump_btn').onclick =
function(){");
            res.append("     var page_size = document.getElementById('pager_jump_
page_size').value;");
            res.append("     if (!/^[1-9]\\d*$/.test(page_size) || page_size < 1 ||
page_size > "+ this.totalPage +"){");
            res.append("         alert('请输入[1-"+this.totalPage +"]之间的页码！');");
            res.append("     }else{");
            res.append("         var submit_url = '" + this.submitUrl + "';");
            res.append("         window.location = submit_url.replace('"+ TAG +"',
page_size);");
            res.append("     }");
            res.append("}");
            res.append("</script>");


        }else{
            res.append("<li><a style=\"background-color:#D4D4D4;\" href=\"#\">总共
<font color='red'>0</font>条记录,当前显示 0-0 条记录。</a> </li>");
        }

        res.append("</ul></nav></p></center>");
        this.getJspContext().getOut().print(res.toString());
    }


    /** 计算中间页码的方法 */
    private void calcPage(StringBuilder str) {
        /** 判断总页数 */
        if (this.totalPage <= 11){
            /** 一次性显示全部页码 */
            for (int i = 1; i <= this.totalPage; i++){
                if (this.pageIndex == i){
                    /** 当前页码 */
                    str.append("<li class=\"active\" ><a href=\"#\">"+i+"</a></li>");
                }else{
                    //2 getUser.action?pageIndex=3?pageSize=14
                    //           3*14 = 42
                    String tempUrl = this.submitUrl.replace(TAG, String.valueOf(i));
                    str.append("<li><a href='"+tempUrl+"'>"+i+"</a></li>");
                }
            }
        }else{
            /** 靠首页近些 */
            if (this.pageIndex <= 8){
                for (int i = 1; i <= 10; i++){
                    if (this.pageIndex == i){
                        /** 当前页码 */
                        str.append("<li class=\"active\" ><a href=\"#\">"+i+"</a></li>");
                    }else{
                        String tempUrl = this.submitUrl.replace(TAG, String.valueOf(i));
                        str.append("<li><a href='"+tempUrl+"'>"+i+"</a></li>");
                    }
                }
                str.append("<li><a href=\"#\">...</a></li>");
                String tempUrl = this.submitUrl.replace(TAG, String.valueOf(this.
```

```java
totalPage));
                str.append("<li><a href='"+tempUrl+"'>"+this.totalPage+"</a></li>");
            }
            /** 靠尾页近些 */
            else if (this.pageIndex + 8 >= this.totalPage){
                String tempUrl = this.submitUrl.replace(TAG, String.valueOf(1));
                str.append("<li><a href='"+tempUrl+"'>1</a></li>");
                str.append("<li><a href=\"#\">...</a></li>");

                for (int i = this.totalPage - 10; i <= this.totalPage; i++){
                    if (this.pageIndex == i){
                        /** 当前页码 */
                        str.append("<li class=\"active\" ><a href=\"#\">"+i+"</a></li>");
                    }else{
                        tempUrl = this.submitUrl.replace(TAG, String.valueOf(i));
                        str.append("<li><a href='"+tempUrl+"'>"+i+"</a></li>");
                    }
                }
            }
            /** 在中间*/
            else{
                String tempUrl = this.submitUrl.replace(TAG, String.valueOf(1));
                str.append("<li><a href='"+tempUrl+"'>1</a></li>");
                str.append("<li><a href=\"#\">...</a></li>");

                for (int i = this.pageIndex - 4; i <= this.pageIndex + 4; i++){
                    if (this.pageIndex == i){
                        /** 当前页码 */
                        str.append("<li class=\"active\" ><a href=\"#\">"+i+"</a></li>");
                    }else{
                        tempUrl = this.submitUrl.replace(TAG, String.valueOf(i));
                        str.append("<li><a href='"+tempUrl+"'>"+i+"</a></li>");
                    }
                }

                str.append("<li><a href=\"#\">...</a></li>");
                tempUrl = this.submitUrl.replace(TAG, String.valueOf(this.totalPage));
                str.append("<li><a href='"+tempUrl+"'>"+this.totalPage+"</a></li>");
            }
        }
    }

    /** setter 方法 */
    public void setPageIndex(int pageIndex) {
        this.pageIndex = pageIndex;
    }
    public void setPageSize(int pageSize) {
        this.pageSize = pageSize;
    }
    public void setRecordCount(int recordCount) {
        this.recordCount = recordCount;
    }
    public void setSubmitUrl(String submitUrl) {
        this.submitUrl = submitUrl;
    }
    public void setStyle(String style) {
        this.style = style;
    }
}
```

要使用 JSP 的标签还需要在 WEB-INF 下增加一个 tld 标签文件。

程序清单：codes/07/oa/webapp/WEB-INF/jsp/pager.tld

```xml
<?xml version="1.0" encoding="utf-8"?>
<taglib xmlns="http://java.sun.com/xml/ns/javaee"
    xmlns:xsi="http://www.w3.org/2001/XMLSchema-instance"
    xsi:schemaLocation="http://java.sun.com/xml/ns/javaee
                    http://java.sun.com/xml/ns/javaee/web-jsptaglibrary_2_1.xsd"
                version="2.1">

    <!-- 描述 自定义标签版本的一种描述 -->
    <description>Pager 1.0 core library</description>
    <!-- 显示的名称 -->
    <display-name>Pager core</display-name>
    <!-- 版本号 -->
    <tlib-version>1.0</tlib-version>
    <!-- 短名称 -->
    <short-name>fkjava</short-name>
    <!-- uri：标签使用者需要导入的包名 -->
    <uri>/pager-tags</uri>

    <!-- 定义一个标签 -->
    <tag>
        <!-- 标签名 -->
        <name>pager</name>
        <!-- 标签处理类 -->
        <tag-class>org.fkit.common.util.pager.PagerTag</tag-class>
        <!-- 设置标签为空 -->
        <body-content>empty</body-content>

        <!-- 定义标签的属性 -->
        <attribute>
            <!-- 属性名 表示分页的第几页 -->
            <name>pageIndex</name>
            <!--required 表示该属性是否必需,true 表示属性必须存在,false 表示属性可以不存在 -->
            <required>true</required>
            <!--rtexprvalue 表示 run time expression value（运行时是否支持表达式），为 true 支持 EL 表达式 -->
            <rtexprvalue>true</rtexprvalue>
        </attribute>

        <!-- 定义标签的属性 -->
        <attribute>
            <!-- 属性名 表示分页标签，每页显示多少条数据 -->
            <name>pageSize</name>
            <!--required 表示该属性是否必需,true 表示属性必须存在,false 表示属性可以不存在 -->
            <required>true</required>
            <!--rtexprvalue 表示 run time expression value（运行时是否支持表达式），为 true 支持 EL 表达式 -->
            <rtexprvalue>true</rtexprvalue>
        </attribute>
        <!-- 定义标签的属性 -->
        <attribute>
            <!-- 属性名 记录分页的总数 -->
            <name>recordCount</name>
            <!--required 表示该属性是否必需,true 表示属性必须存在,false 表示属性可以不存在 -->
            <required>true</required>
            <!--rtexprvalue 表示 run time expression value（运行时是否支持表达式），为 true 支持 EL 表达式 -->
            <rtexprvalue>true</rtexprvalue>
        </attribute>
        <!-- 定义标签的属性 -->
```

```xml
            <attribute>
                <!-- 属性名 -->
                <name>submitUrl</name>
                <!--required 表示该属性是否必需,true 表示属性必须存在,false 表示属性可以不存在 -->
                    <required>true</required>
                <!--rtexprvalue 表示 run time expression value（运行时是否支持表达式），为
true 支持 EL 表达式 -->
                <rtexprvalue>true</rtexprvalue>
            </attribute>
            <!-- 定义标签的属性 -->
            <attribute>
                <!-- 属性名 -->
                <name>style</name>
                <!- required 表示该属性是否必需, true 表示属性必须存在, false 表示属性可以不存在 -->
                    <required>false</required>
                <!--rtexprvalue 表示 run time expression value（运行时是否支持表达式），为
true 支持 EL 表达式 -->
                <rtexprvalue>true</rtexprvalue>
            </attribute>
    </tag>
</taglib>
```

接下来是 IdentityService 接口的源代码。

程序清单：codes/07/oa/src/main/java/org/fkit/oa/identity/service/IdentityService

```java
import java.util.List;
import java.util.Map;
import javax.servlet.http.HttpSession;
import org.fkit.common.util.pager.PageModel;
import org.fkit.oa.identity.domain.Dept;
import org.fkit.oa.identity.domain.Module;
import org.fkit.oa.identity.domain.Role;
import org.fkit.oa.identity.domain.User;
import org.fkit.oa.identity.dto.UserModule;
import org.fkit.oa.identity.vo.TreeData;

public interface IdentityService {

    /**
     * @return 查询所有部门
     */
    List<Dept> getAllDepts() ;

    /**
     * 异步登录的业务层接口方法
     * @param params
     * @return
     */
    Map<String, Object> login(Map<String, Object> params);

    /**
     * 根据用户的主键查询用户信息,包含了延迟加载的部门和职位信息
     * @param userId
     * @return
     */
    User getUserById(String userId);

    /**
     * 修改自己
     * @param user
     */
    void updateSelf(User user,HttpSession session);
```

```java
/**
 * @return 异步加载部门与职位的 JSON 字符串信息，写回页面
 */
Map<String, Object> getAllDeptsAndJobsAjax();

/**
 * 分页查询用户信息
 *
 * @param user
 * @param pageModel
 * @return
 */
List<User> getUsersByPager(User user, PageModel pageModel);

/**
 * 批量删除用户
 * @param ids
 */
void deleteUserByUserIds(String ids);

/**
 * 校验用户是否已经被注册
 * @param userId
 * @return
 */
String isUserValidAjax(String userId);

/**
 *
 * @param user
 */
void addUser(User user);

/**
 * 根据 userId 修改用户信息
 * @param user
 */
void updateUser(User user);

/**
 * 激活用户
 * @param user
 */
void activeUser(User user);

/**
 * 加载所有的模块树
 * @return
 */
List<TreeData> loadAllModuleTrees();

/**
 * 根据父节点查询所有的子模块
 * @param parentCode
 * @return
 */
List<Module> getModulesByParent(String parentCode,PageModel pageModel);
/**
 * 根据父节点查询所有的子模块
 * 不分页
 * @param parentCode
 * @return
 */
```

```java
    List<Module> getModulesByParent(String parentCode);

    /**
     * 批量删除菜单
     * @param ids
     */
    void deleteModules(String ids);

    /**
     * 为当前父节点菜单添加子节点模块
     * @param parentCode
     * @param module
     */
    void addModule(String parentCode, Module module);

    /**
     * 根据编号查询模块信息
     * @param code
     * @return
     */
    Module getModuleByCode(String code);

    /**
     * 修改模块
     * @param module
     */
    void updateModule(Module module);

    /**
     * 分页查询角色信息
     * @param pageModel
     * @return
     */
    List<Role> getRoleByPager(PageModel pageModel);

    /**
     * 添加角色
     * @param role
     */
    void addRole(Role role);

    /**
     * 批量删除角色
     * @param ids
     */
    void deleteRole(String ids);

    /**
     * 根据id查询角色
     * @param id
     * @return
     */
    Role getRoleById(Long id);

    /**
     * 修改角色
     * @param role
     */
    void updateRole(Role role);

    /**
     * 分页查询属于这个角色的用户信息
     * @param role
```

```java
 *  @param pageModel
 *  @return
 */
List<User> selectRoleUser(Role role, PageModel pageModel);

/**
 * 查询不属于某个角色的用户
 * @param role
 * @param pageModel
 * @return
 */
List<User> selectNotRoleUser(Role role, PageModel pageModel);

/**
 * 给用户绑定角色
 * @param role
 * @param ids
 */
void bindUser(Role role, String ids);

/**
 * 给用户解绑角色
 * @param role
 * @param ids
 */
void unBindUser(Role role, String ids);

/**
 * 查询当前角色在当前模块下拥有的操作权限编号
 * @param role
 * @param parentCode
 * @return
 */
List<String> getRoleModuleOperasCodes(Role role, String parentCode);

/**
 * 给角色绑定某个模块下的操作权限
 * @param codes
 * @param role
 * @param parentCode
 */
void bindPopedom(String codes, Role role, String parentCode);

/**
 * 查询当前用户的权限模块
 * @return
 */
List<UserModule> getUserPopedomModules();
}
```

IdentityService 接口中包含本系统所有业务逻辑方法的定义，接下来是这些业务逻辑方法的实现。

程序清单：codes/07/oa/src/main/java/org/fkit/oa/identity/service/impl/IdentityServiceImpl

```java
import java.util.ArrayList;
import java.util.Date;
import java.util.HashMap;
import java.util.LinkedHashMap;
import java.util.List;
import java.util.Map;
import java.util.Map.Entry;
import javax.persistence.criteria.CriteriaBuilder;
```

```java
import javax.persistence.criteria.CriteriaQuery;
import javax.persistence.criteria.JoinType;
import javax.persistence.criteria.Path;
import javax.persistence.criteria.Predicate;
import javax.persistence.criteria.Root;
import javax.servlet.http.HttpSession;
import org.fkit.common.util.CommonContants;
import org.fkit.common.util.pager.PageModel;
import org.fkit.oa.identity.domain.Dept;
import org.fkit.oa.identity.domain.Module;
import org.fkit.oa.identity.domain.Popedom;
import org.fkit.oa.identity.domain.Role;
import org.fkit.oa.identity.domain.User;
import org.fkit.oa.identity.dto.UserModule;
import org.fkit.oa.identity.repository.DeptRepository;
import org.fkit.oa.identity.repository.JobRepository;
import org.fkit.oa.identity.repository.ModuleRepository;
import org.fkit.oa.identity.repository.PopedomRepository;
import org.fkit.oa.identity.repository.RoleRepository;
import org.fkit.oa.identity.repository.UserRepository;
import org.fkit.oa.identity.service.IdentityService;
import org.fkit.oa.identity.vo.TreeData;
import org.fkit.oa.util.OaContants;
import org.fkit.oa.util.OaException;
import org.fkit.oa.util.UserHolder;
import org.springframework.beans.factory.annotation.Autowired;
import org.springframework.beans.factory.annotation.Qualifier;
import org.springframework.data.domain.Page;
import org.springframework.data.domain.PageRequest;
import org.springframework.data.domain.Pageable;
import org.springframework.data.domain.Sort;
import org.springframework.data.jpa.domain.Specification;
import org.springframework.stereotype.Service;
import org.springframework.transaction.annotation.Transactional;
import org.springframework.util.StringUtils;
/**
 * 系统管理的业务层实现类
 */
@Service("identityService")  // 配置业务层的bean
//@Transactional(readOnly=false,rollbackFor=java.lang.RuntimeException.class)
@Transactional(readOnly=true)
public class IdentityServiceImpl implements IdentityService {

    /** 定义Repository对象 */
    @Autowired
    @Qualifier("deptRepository")
    private DeptRepository deptRepository;

    @Autowired
    @Qualifier("jobRepository")
    private JobRepository jobRepository;

    @Autowired
    @Qualifier("userRepository")
    private UserRepository userRepository;

    @Autowired
    @Qualifier("moduleRepository")
    private ModuleRepository moduleRepository;

    @Autowired
    @Qualifier("roleRepository")
    private RoleRepository roleRepository;
```

```java
    @Autowired
    @Qualifier("popedomRepository")
    private PopedomRepository popedomRepository;

    @Transactional(readOnly=true)
    @Override
    public List<Dept> getAllDepts(){
        // 思考异常的处理
        try {
            List<Dept> depts = deptRepository.findAll();
            // 获取延迟加载的属性，会话此时并没有关闭
            for(Dept dept : depts){
                if(dept.getCreater()!=null)dept.getCreater().getName();
                if(dept.getModifier()!=null)dept.getModifier().getName();
            }
            return depts;
        } catch (Exception e) {
            throw new OaException("查询部门失败了", e);
        }
    }

    @Override
    public Map<String , Object> login(Map<String, Object> params) {
        Map<String , Object> result = new HashMap<>();
        try {
            /** 处理登录的业务逻辑   */
            /** 1.参数非空校验   */
            String userId = (String) params.get("userId");
            String passWord = (String) params.get("passWord");
            String vcode = (String) params.get("vcode");
            HttpSession session = (HttpSession) params.get("session");
            // userId!=null&&!userId.equals("")
            if(StringUtils.isEmpty(userId) || StringUtils.isEmpty(passWord)
                    || StringUtils.isEmpty(vcode) ){
                /** 参数有为空的 */
                result.put("status", 0);
                result.put("tip", "参数有为空的");
            }else{
                /** 参数不为空   */
                /** 校验验证码是否正确
                 *  获取 session 中当前用户对应的验证码
                 * */
                String sysCode = (String) session.getAttribute(CommonContants.VERIFY_SESSION);
                if(vcode.equalsIgnoreCase(sysCode)){
                    /** 验证码正确了   */
                    /** 根据登录的用户名查询用户：判断登录名是否存在   */
                    User user = getUserById(userId);
                    if(user!=null){
                        /** 登录名存在    */
                        /** 判断密码 */
                        if(user.getPassWord().equals(passWord)){
                            /** 判断用户是否已经被激活 */
                            if(user.getStatus() == 1){
                                /** 登录成功   */
                                /** 1.把登录成功的用户放入当前用户的 Session 会话中   */
                                session.setAttribute(OaContants.USER_SESSION, user);
                                result.put("status",1);
                                result.put("tip", "登录成功");
                                /** 把登录成功的用户存入 UserHolder*/
                                UserHolder.addCurrentUser(user);
                                /** 2.当用户登录系统，应该立即查询该用户所拥有
```

```java
                                  的全部操作权限 --> 存入当前用户的Session会话中   */
                            Map<String,List<String>>  userAllOperasPopedomUrls =
getUserAllOperasPopedomUrls();
                                session.setAttribute(OaContants.USER_ALL_OPERAS_
POPEDOM_URLS, userAllOperasPopedomUrls);
                        }else{
                            result.put("status", 5);
                            result.put("tip", "您的账号未被激活,请联系管理员激活!");
                        }
                    }else{
                        /** 密码错误      */
                        result.put("status", 2);
                        result.put("tip", "密码错误");
                    }
                }else{
                    /** 登录名不存在   */
                    result.put("status", 3);
                    result.put("tip", "没有该账户");
                }
            }else{
                /** 验证码不正确 */
                result.put("status", 4);
                result.put("tip", "验证码不正确");
            }
        }
        return result;
    } catch (Exception e) {
        throw new OaException("异步登录业务层抛出异常了", e);
    }
}

private Map<String, List<String>> getUserAllOperasPopedomUrls() {
    try {
        /** 查询用户所拥有的所有操作权限编号
         *  [000100010001,000100010002]
         * */
        List<String> userAllPopedomOperasCodes = popedomRepository.
getUserPopedomOperasCodes(UserHolder.getCurrentUser().getUserId());
        if(userAllPopedomOperasCodes!=null && userAllPopedomOperasCodes.size()>0 ){
            Map<String, List<String>> userAllOperasPopedomUrls = new HashMap<>();
            String moduleUrl = "" ;
            List<String> moduleOperaUrls = null;
            for(String operaCode : userAllPopedomOperasCodes){
                /** 先得到模块的编号   */
                String parentModuleCode = operaCode.substring(0, operaCode.length()
- OaContants.CODE_LEN);
                /** 父模块地址 */
                moduleUrl = getModuleByCode(parentModuleCode).getUrl();
                /** 判断map集合中是否已经存在该父模块地址 */
                if(!userAllOperasPopedomUrls.containsKey(moduleUrl)){
                    moduleOperaUrls = new ArrayList<String>();
                    userAllOperasPopedomUrls.put(moduleUrl, moduleOperaUrls);
                }
                moduleOperaUrls.add(getModuleByCode(operaCode).getUrl());
            }
            return userAllOperasPopedomUrls;
        }
        return null;
    } catch (Exception e) {
        throw new OaException("登录查询用户的操作权限出现异常", e);
```

```java
        }
    }

    public User getUserById(String userId) {
        try {
            User user = userRepository.findOne(userId);
            if(user != null){
                // 获取延迟加载的属性
                if(user.getDept()!=null)user.getDept().getName();
                if(user.getJob()!=null)user.getJob().getName();
                if(user.getCreator()!=null)user.getCreator().getName();
                if(user.getModifier()!=null)user.getModifier().getName();
                if(user.getChecker()!=null)user.getChecker().getName();
                return user;
            }
            return null ;
        } catch (Exception e) {
            throw new OaException("查询用户失败了", e);
        }
    }

    @Transactional
    @Override
    public void updateSelf(User user,HttpSession session) {
        try {
            /** 1.持久化修改    */
            User sessionUser = userRepository.findOne(user.getUserId());
            sessionUser.setModifyDate(new Date());
            sessionUser.setModifier(user);
            sessionUser.setName(user.getName());
            sessionUser.setEmail(user.getEmail());
            sessionUser.setTel(user.getTel());
            sessionUser.setPhone(user.getPhone());
            sessionUser.setQuestion(user.getQuestion());
            sessionUser.setAnswer(user.getAnswer());
            sessionUser.setQqNum(user.getQqNum());
            // user = sessionUser;
            // 获取延迟加载的属性
            if(sessionUser.getDept()!=null)sessionUser.getDept().getName();
            if(sessionUser.getJob()!=null)sessionUser.getJob().getName();
            session.setAttribute(OaContants.USER_SESSION, sessionUser);
        } catch (Exception e) {
            throw new OaException("修改用户失败了", e);
        }
    }

    @Override
    public Map<String , Object > getAllDeptsAndJobsAjax() {
        try {
            /** 1.定义一个 Map 对象封装最终查询出来的部门信息和职位信息 */
            Map<String , Object > deptJobDatas = new HashMap<>();
            /** 查询部门: id name
             * deptsList = [ {id=1, name="开发部"} , {id=2, name="销售部"} ]
             * */
            List<Map<String , Object>> deptsList = deptRepository.findDepts();

            /** 查询职位: id name
             * jobLists = [ {id=1, name="java"} , {id=2, name="咨询师"} ]
             * */
            List<Map<String , Object>> jobLists = jobRepository.findJobs();
```

```java
            deptJobDatas.put("depts", deptsList);
            deptJobDatas.put("jobs", jobLists);
            
            return deptJobDatas;
        } catch (Exception e) {
            throw new OaException("查询部门与职位信息异常了", e);
        }
    }
    
    @Override
    public List<User> getUsersByPager(User user, PageModel pageModel) {
        try {
            Page<User> usersPager = userRepository.findAll(new Specification<User>() {
                @Override
                public Predicate toPredicate(Root<User> root, CriteriaQuery<?> query,
                        CriteriaBuilder cb) {
                    // 本集合用于封装查询条件
                    List<Predicate> predicates = new ArrayList<Predicate>();
                    if(user!=null){
                        /** 是否传入用于查询的姓名 */
                        if(!StringUtils.isEmpty(user.getName())){
                            predicates.add(cb.like(root.<String> get("name"),"%" + user.getName() + "%"));
                        }
                        /** 是否传入用于查询的手机号码 */
                        if(!StringUtils.isEmpty(user.getPhone())){
                            predicates.add(cb.like(root.<String> get("phone"),"%" + user.getPhone() + "%"));
                        }
                        /** 是否传入用于查询的部门 */
                        if(user.getDept()!=null && user.getDept().getId()!=null && user.getDept().getId()!= 0 ){
                            root.join("dept", JoinType.INNER);
                            Path<Long> d_id = root.get("dept").get("id");
                            predicates.add(cb.equal(d_id, user.getDept().getId()));
                        }
                        if(user.getJob()!=null && !StringUtils.isEmpty(user.getJob().getCode())
                                && !user.getJob().getCode().equals("0")){
                            root.join("job", JoinType.INNER);
                            Path<String> j_id = root.get("job").get("code");
                            predicates.add(cb.equal(j_id, user.getJob().getCode()));
                        }
                    }
                    return query.where(predicates.toArray(new Predicate[predicates.size()])).getRestriction();
                }
            }, PageRequest.of(pageModel.getPageIndex() - 1, pageModel.getPageSize()));
            pageModel.setRecordCount(usersPager.getTotalElements());
            /** 获取每个用户的延迟加载属性 */
            List<User> users = usersPager.getContent();
            for(User u : users){
                if(u.getDept()!=null)u.getDept().getName();
                if(u.getJob()!=null)u.getJob().getName();
                if(u.getChecker()!=null)u.getChecker().getName();
            }
            return users;
        } catch (Exception e) {
            throw new OaException("查询用户信息异常了", e);
        }
    }
    
    @Transactional
    @Override
```

```java
public void deleteUserByUserIds(String ids) {
    try {
        List<User> users = new ArrayList<User>();
        for(String id : ids.split(",")){
            User user = new User() ;
            user.setUserId(id);
            users.add(user);
        }
        userRepository.deleteInBatch(users);
    } catch (Exception e) {
        throw new OaException("删除用户信息异常了", e);
    }
}

@Override
public String isUserValidAjax(String userId) {
    try {
        User user = userRepository.findOne(userId);
        return user==null?"success":"error";
    } catch (Exception e) {
        throw new OaException("校验用户登录名是否注册异常了", e);
    }
}

@Transactional
@Override
public void addUser(User user) {
    try {
        user.setCreateDate(new Date());
        user.setCreater(UserHolder.getCurrentUser());
        userRepository.save(user);
    } catch (Exception e) {
        throw new OaException("添加用户信息异常了", e);
    }
}

@Transactional
@Override
public void updateUser(User user) {
    try {
        /** 1.持久化修改     */
        User sessionUser = userRepository.findOne(user.getUserId());
        sessionUser.setModifyDate(new Date());
        sessionUser.setModifier(UserHolder.getCurrentUser());
        sessionUser.setPassWord(user.getPassWord());
        sessionUser.setName(user.getName());
        sessionUser.setDept(user.getDept());
        sessionUser.setJob(user.getJob());
        sessionUser.setEmail(user.getEmail());
        sessionUser.setSex(user.getSex());
        sessionUser.setTel(user.getTel());
        sessionUser.setPhone(user.getPhone());
        sessionUser.setQuestion(user.getQuestion());
        sessionUser.setAnswer(user.getAnswer());
        sessionUser.setQqNum(user.getQqNum());
    } catch (Exception e) {
        throw new OaException("修改用户失败了", e);
    }
}

@Transactional
@Override
public void activeUser(User user) {
```

```java
            try {
                User sessionUser = userRepository.findOne(user.getUserId());
                sessionUser.setCheckDate(new Date());
                sessionUser.setChecker(UserHolder.getCurrentUser());
                sessionUser.setStatus(user.getStatus());
            } catch (Exception e) {
                throw new OaException("激活用户失败了", e);
            }
        }

        @Override
        public List<TreeData> loadAllModuleTrees() {
            try {
                /** 查询所有模块信息   */
                List<Module> modules = moduleRepository.findAll();
                /** 拼装成 dtree 需要的树节点  */
                List<TreeData> treeDatas = new ArrayList<>();
                for(Module m : modules){
                    TreeData data = new TreeData();
                    data.setId(m.getCode());
                    data.setName(m.getName());
                    // 长度为4的编号的父节点是0
                    // 其余节点的父节点是从开始位置一直截取到总长度减去步长的位置
                    String pid = m.getCode().length()==OaContants.CODE_LEN?"0":
m.getCode().substring(0, m.getCode().length()-OaContants.CODE_LEN);
                    data.setPid(pid);
                    treeDatas.add(data);
                }
                return treeDatas;
            } catch (Exception e) {
                throw new OaException("加载模块树异常", e);
            }
        }

        @Override
        public List<Module> getModulesByParent(String parentCode,PageModel pageModel) {
            try {

                parentCode = parentCode==null?"":parentCode;
                List<Object> values = new ArrayList<>();
                values.add(parentCode+"%");
                values.add(parentCode.length()+OaContants.CODE_LEN);
//              // 子节点的编号的长度是父节点编号长度+步长
//              // 子节点前几位的编号必须与父节点编码一致
                Page<Module> modulesPager = moduleRepository.findAll(new Specification<Module>() {
                    @Override
                    public Predicate toPredicate(Root<Module> root, CriteriaQuery<?> query,
                        CriteriaBuilder cb) {
                        // 本集合用于封装查询条件
                        List<Predicate> predicates = new ArrayList<Predicate>();
                        predicates.add(cb.like(root.<String> get("code"),values.get(0)+""));
                        predicates.add(cb.equal(cb.length(root.<String> get("code")),
values.get(1)));
                        return query.where(predicates.toArray(new Predicate[predicates.size()])).getRestriction();
                    }
                }, PageRequest.of(pageModel.getPageIndex() - 1, pageModel.getPageSize()));
                pageModel.setRecordCount(modulesPager.getTotalElements());
                /** 取每个用户的延迟加载属性 */
                List<Module> sonModules = modulesPager.getContent();
                for(Module m : sonModules){
```

```java
                if(m.getCreater()!=null)m.getCreater().getName();
                if(m.getModifier()!=null)m.getModifier().getName();
            }
            return sonModules;
        } catch (Exception e) {
            throw new OaException("查询子模块异常", e);
        }
    }

    @Override
    public List<Module> getModulesByParent(String parentCode) {
        try {

            parentCode = parentCode==null?"":parentCode;
            List<Module> sonModules = moduleRepository.findModules(parentCode+"%" , parentCode.length()+OaContants.CODE_LEN);
            for(Module m : sonModules){
                if(m.getCreater()!=null)m.getCreater().getName();
                if(m.getModifier()!=null)m.getModifier().getName();
            }
            return sonModules;
        } catch (Exception e) {
            throw new OaException("查询子模块异常", e);
        }
    }

    @Transactional
    @Override
    public void deleteModules(String ids) {
        try {
            for(String code  : ids.split(",")){
                moduleRepository.setCode(code);
            }
        } catch (Exception e) {
            throw new OaException("批量删除菜单异常", e);
        }

    }

    @Transactional
    @Override
    public void addModule(String parentCode, Module module) {
        try {

            /** 维护编号:通用工具类(给你一个父节点，给你一张表，给你那个字段，
             *   找出该字段该父节点下的下一个子节点的编号）  */
            module.setCode(getNextSonCode(parentCode, OaContants.CODE_LEN));
            module.setCreateDate(new Date());
            module.setCreater(UserHolder.getCurrentUser());
            moduleRepository.save(module);
        } catch (Exception e) {
            throw new OaException("添加子菜单异常", e);
        }

    }

    public String getNextSonCode(String parentCode,int codeLen) throws Exception {
        /** 判断父节点是否为 null */
        parentCode =  parentCode==null?"":parentCode;
        /** 1.查询当前父节点下的最大子节点编号 */
        String maxSonCode = moduleRepository.findUniqueEntity(parentCode+"%" , parentCode.length()+codeLen);
        System.out.println("当前最大子节点编号是: maxSonCode---->"+maxSonCode);
```

```java
            String nextSonCode = "";  // 保存最终的下一个子节点编号
            /** 2.判断最大子节点编号是否存在，因为极有可能父节点此时一个子节点都没有 */
            if(StringUtils.isEmpty(maxSonCode)){
                /** 子节点编号不存在 */
                String preSuffix = "" ;  // 0 需要拼接多少个 0
                for(int i = 0 ; i < codeLen - 1; i++ ){
                    preSuffix+="0";
                }
                nextSonCode = parentCode+preSuffix+1;
            }else{
                /** 子节点编号存在 --> 000100010005 */
                /** 截取当前子节点编号的步长 */
                String currentMaxSonCode = maxSonCode.substring(parentCode.length());
                /** 得到子节点步长编号的整型形式 */
                int maxSonCodeInt = Integer.valueOf(currentMaxSonCode);
                maxSonCodeInt++;
                /** 判断编号是否越界 */
                if((maxSonCodeInt+"").length() > codeLen){
                    throw new OaException("编号越界了！");
                }else{
                    /** 没有越界 */
                    String preSuffix = "" ;  // 0 需要拼接多少个 0
                    for(int i = 0 ; i< codeLen-(maxSonCodeInt+"").length() ; i++){
                        preSuffix+="0";
                    }
                    
                    nextSonCode = parentCode+preSuffix+maxSonCodeInt;
                }
            }
            return nextSonCode;
    }
    
    @Override
    public Module getModuleByCode(String code) {
        try {
            return moduleRepository.findOne(code);
        } catch (Exception e) {
            throw new OaException("查询模块异常", e);
        }
    }
    
    @Transactional
    @Override
    public void updateModule(Module module) {
        try {
            Module sessionModule = moduleRepository.findOne(module.getCode());
            sessionModule.setModifier(UserHolder.getCurrentUser());
            sessionModule.setModifyDate(new Date());
            sessionModule.setName(module.getName());
            sessionModule.setRemark(module.getRemark());
            sessionModule.setUrl(module.getUrl());
        } catch (Exception e) {
            throw new OaException("修改模块异常", e);
        }
        
    }
    
    @Override
    public List<Role> getRoleByPager(PageModel pageModel) {
        try {
            // 指定排序参数对象：根据 id，进行降序查询
            Sort.Order order = new Sort.Order(Sort.Direction.ASC, "id");
            Sort sort = new Sort(order);
```

```java
            /**
             * 封装分页实体
             * 参数1：pageIndex 表示当前查询第几页（默认从 0 开始，0 表示第 1 页）
             * 参数2：表示每页展示多少数据，现在设置每页展示 2 条数据
             * 参数3：封装排序对象，根据该对象的参数指定根据 id 降序查询
             */
            Pageable page = PageRequest.of(pageModel.getPageIndex() - 1, pageModel.getPageSize(), sort);
            Page<Role> rolePager = roleRepository.findAll(page);
            pageModel.setRecordCount(rolePager.getTotalElements());
            /** 获取每个用户的延迟加载属性 */
            List<Role> roles = rolePager.getContent();
            for(Role r : roles){
                if(r.getModifier()!=null)r.getModifier().getName();
                if(r.getCreater()!=null)r.getCreater().getName();
            }
            return roles;
        } catch (Exception e) {
            throw new OaException("查询角色异常", e);
        }
    }

    @Transactional
    @Override
    public void addRole(Role role) {
        try {
            role.setCreateDate(new Date());
            role.setCreater(UserHolder.getCurrentUser());
            roleRepository.save(role);
        } catch (Exception e) {
            throw new OaException("添加角色异常", e);
        }

    }

    @Transactional
    @Override
    public void deleteRole(String ids) {
        try {
            List<Role> roles = new ArrayList<Role>();
            for(String id : ids.split(",")){
                Role role = new Role() ;
                role.setId(Long.valueOf(id));
            }
            roleRepository.deleteInBatch(roles);

        } catch (Exception e) {
            throw new OaException("批量删除角色异常", e);
        }
    }

    @Override
    public Role getRoleById(Long id) {
        try {
            return roleRepository.findOne(id);
        } catch (Exception e) {
            throw new OaException("根据 id 查询角色异常", e);
        }
    }

    @Transactional
    @Override
    public void updateRole(Role role) {
        try {
            Role r = roleRepository.findOne(role.getId());
```

```java
                r.setName(role.getName());
                r.setRemark(role.getRemark());
                r.setModifier(UserHolder.getCurrentUser());
                r.setModifyDate(new Date());
            } catch (Exception e) {
                throw new OaException("根据id修改角色异常", e);
            }
        }


        @Override
        public List<User> selectRoleUser(Role role, PageModel pageModel) {
            try {
                Page<User> usersPager = userRepository.findAll(new Specification<User>() {
                    @Override
                    public Predicate toPredicate(Root<User> root, CriteriaQuery<?> query,
                            CriteriaBuilder cb) {
                        // 本集合用于封装查询条件
                        List<Predicate> predicates = new ArrayList<>();
                        List<String> userIds = userRepository.findRoleUsers(role.getId());
                        predicates.add(root.<String>get("userId").in(userIds));
                        return query.where(predicates.toArray(new Predicate[predicates.size()])).getRestriction();
                    }
                }, PageRequest.of(pageModel.getPageIndex() - 1, pageModel.getPageSize()));
                pageModel.setRecordCount(usersPager.getTotalElements());
                List<User> users = usersPager.getContent();
                for(User u : users){
                    if(u.getDept()!=null)u.getDept().getName();
                    if(u.getJob()!=null)u.getJob().getName();
                    if(u.getChecker()!=null)u.getChecker().getName();
                }
                return users;
            } catch (Exception e) {
                throw new OaException("查询属于角色下的用户信息异常", e);
            }
        }



        @Override
        public List<User> selectNotRoleUser(Role role, PageModel pageModel) {
            try {

                Page<User> usersPager = userRepository.findAll(new Specification<User>() {
                    @Override
                    public Predicate toPredicate(Root<User> root, CriteriaQuery<?> query,
                            CriteriaBuilder cb) {
                        // 本集合用于封装查询条件
                        List<Predicate> predicates = new ArrayList<Predicate>();
                        // 先查询出不属于这个角色的用户
                        List<String> userId = userRepository.getRolesUsers(role.getId());
                        predicates.add(root.<String>get("userId").in(userId));
                        return query.where(predicates.toArray(new Predicate[predicates.size()])).getRestriction();
                    }
                }, PageRequest.of(pageModel.getPageIndex() - 1, pageModel.getPageSize()));
                pageModel.setRecordCount(usersPager.getTotalElements());
                List<User> users = usersPager.getContent();

                for(User u : users){
                    if(u.getDept()!=null)u.getDept().getName();
                    if(u.getJob()!=null)u.getJob().getName();
                    if(u.getChecker()!=null)u.getChecker().getName();
```

```java
            return users;
        } catch (Exception e) {
            throw new OaException("查询不属于这个角色的用户信息异常", e);
        }
    }

    @Transactional
    @Override
    public void bindUser(Role role, String ids) {
        try {
            /** 给角色绑定一批用户 */
            /** 1.先查询该角色 */
            Role session = roleRepository.findOne(role.getId());
            /** 2.给角色的users 添加需要绑定的用户 */
            for(String userId : ids.split(",")){
                User user = userRepository.findOne(userId);
                session.getUsers().add(user);
            }

        } catch (Exception e) {
            throw new OaException("绑定角色下的用户异常", e);
        }

    }

    @Transactional
    @Override
    public void unBindUser(Role role, String ids) {
        try {
            /** 给角色绑定一批用户 */
            /** 1.先查询该角色 */
            Role session = roleRepository.findOne(role.getId());
            /** 2.给角色的users 添加需要绑定的用户 */
            for(String userId : ids.split(",")){
                User user = userRepository.findOne(userId);
                session.getUsers().remove(user);
                // User user = new user("ligang",32)
                // User user1 = new user("ligang",32)
            }

        } catch (Exception e) {
            throw new OaException("绑定角色下的用户异常", e);
        }

    }

    @Override
    public List<String> getRoleModuleOperasCodes(Role role, String parentCode) {
        try {
            List<String> roleModuleOperasCodes = popedomRepository.findByIdAndParentCode(role.getId(),parentCode);
            return roleModuleOperasCodes;
        } catch (Exception e) {
            throw new OaException("查询当前角色在当前模块下拥有的操作权限编号异常", e);
        }
    }

    @Transactional
    @Override
    public void bindPopedom(String codes, Role role, String parentCode) {
        try {
            /** 1.先清空此角色在此模块下的所有操作权限 */
```

```java
            popedomRepository.setByIdAndParentCode(role.getId(),parentCode);
            /** 2.更新角色模块权限 */
            if(!StringUtils.isEmpty(codes)){
                Module parent = getModuleByCode(parentCode);
                /** 添加一些更新的权限 */
                for(String code : codes.split(",")){
                    /** 创建一个权限对象 */
                    Popedom popedom = new Popedom();
                    popedom.setRole(role);
                    popedom.setModule(parent);
                    popedom.setOpera(getModuleByCode(code));
                    popedom.setCreateDate(new Date());
                    popedom.setCreater(UserHolder.getCurrentUser());
                    popedomRepository.save(popedom);
                }
            }
        } catch (Exception e) {
            throw new OaException("给角色绑定某个模块的操作权限异常", e);
        }
    }

    @Override
    public List<UserModule> getUserPopedomModules() {
        try {
            /**查询当前用户的权限模块：先查用户所有的角色，再查这些角色拥有的所有权限模块 */
            List<String> popedomModuleCodes = popedomRepository.getUserPopedomModuleCodes(UserHolder.getCurrentUser().getUserId());
            if(popedomModuleCodes!=null && popedomModuleCodes.size()>0){

                /** 定义一个Map集合用于保存用户的权限模块
                 * Map<Module,List<Module>>
                 * {系统管理=[用户管理,角色管理] , 假期模块=[查询信息,用户请假]}
                 */
                Map<Module,List<Module>> userModulesMap = new LinkedHashMap<>();
                Module fistModule = null ;
                List<Module> secondModules = null ;
                for(String moduleCode : popedomModuleCodes){
                    /** 截取当前模块的一级模块编号 */
                    String fistCode = moduleCode.substring(0, OaContants.CODE_LEN);
                    /** 查询一级模块对象 */
                    fistModule = getModuleByCode(fistCode);
                    fistModule.setName(fistModule.getName().replaceAll("-", ""));
                    /**如果Map集合中没有包含当前一级模块的key,说明是第一次添加一级模块 */
                    if(!userModulesMap.containsKey(fistModule)){
                        secondModules = new ArrayList<Module>();
                        userModulesMap.put(fistModule, secondModules);
                    }
                    Module secondModule = getModuleByCode(moduleCode);
                    secondModule.setName(secondModule.getName().replaceAll("-", ""));
                    secondModules.add(secondModule);
                }

                List<UserModule> userModules = new ArrayList<>();
                /**
                 * 遍历Map集合
                 */
                for(Entry<Module, List<Module>> entry : userModulesMap.entrySet()){
                    Module key = entry.getKey();
                    List<Module> value = entry.getValue();
                    UserModule userModule = new UserModule();
                    userModule.setFirstModule(key);
                    userModule.setSecondModules(value);
```

```
                    userModules.add(userModule);
                }
                return userModules;
            }
            return null;
        } catch (Exception e) {
            throw new OaException("查询当前用户的权限模块", e);
        }
    }
}
```

IdentityServiceImpl 类中实现了服务接口 IdentityService 中定义的所有业务逻辑方法，在 IdentityServiceImpl 类中使用了注解。

> @Transactional(readOnly=true)：表示该类需要 Spring 加入事务，默认属性 Propagation. REQUIRED 指有事务就处于当前事务中，没有事务就创建一个事务；isolation=Isolation. DEFAULT 属性表示使用事务数据库的默认隔离级别。
> @Service("identityService")：将该类配置成一个 Spring 的 bean，标识符是 identityService。
> @Autowired 和@Qualifier：在 IdentityServiceImpl 类中业务方法的实现依赖于 Repository 组件，@Autowired 注解默认使用 byType，@Qualifier("deptRepository")注解使用 byName，自动装配将 6 个持久层的 Repository 注入给 IdentityServiceImpl 类对应依赖的 Repository 组件。

▶▶ 7.5.3 事务管理

与所有的 Java EE 应用类似，本系统的事务管理负责管理业务逻辑组件里的业务逻辑方法，只有对业务逻辑方法添加事务管理才有实际意义，对于单个 Repository 方法（基本的 CRUD 方法）增加事务管理是没有太大实际意义的。

使用 Spring Boot，不再需要在 XML 中对 transactionManager 进行显式的配置，一切都交给 Spring Boot 自动完成。开发中只需在 Java 类中使用@Transactional 注解给 Spring 的 bean 添加事务管理，系统可以非常方便地为业务逻辑组件配置事务管理功能。

7.6 实现 Web 层

前面部分已经实现了本应用的所有中间层内容，系统的所有业务逻辑组件也都部署在 Spring 容器中，接下来应该为应用实现 Web 层了。通常而言，系统的控制器和 JSP 在一起设计。因为当 JSP 页面发出请求后，该请求被控制器接收，然后控制器负责调用业务逻辑组件来处理请求。从这个意义上来说，控制器是 JSP 页面和业务逻辑组件之间的纽带。

▶▶ 7.6.1 控制器

当控制器接收到用户请求后，控制器并不会处理用户请求，只是将用户的请求参数解析处理，然后调用业务逻辑方法来处理用户请求；当请求处理完成后，控制器负责将处理结果通过 JSP 页面呈现给用户。

对于使用 Spring MVC 的应用而言，控制器实际上由两个部分组成：系统的核心控制器 DispatcherServlet 和业务控制器 Controller。使用 Spring Boot，不需要再在 web.xml 中对 DispatcherServlet 进行显式配置，开发者需要关注的就是业务控制器 Controller。

接下来介绍控制层的实现。

7.6.2 系统登录

本系统的所有 JSP 页面都放在 webapp/WEB-INF/jsp 目录下，登录页面是 login.jsp 页面，用户提交登录请求后，用户输入的登录名、密码被异步提交到 LoginController 的 login 方法，LoginController 将会根据请求参数进行用户验证，并将 JSON 数据返回到客户端，客户端根据返回的结果，决定呈现哪个视图资源。

程序清单：codes/07/oa/webapp/WEB-INF/jsp/login.jsp

```jsp
<%@ page language="java" contentType="text/html; charset=UTF-8"
    pageEncoding="UTF-8"%>
<!DOCTYPE html>
<html lang="en">
<head>
    <meta charset="utf-8">
    <meta name="viewport" content="width=device-width, initial-scale=1">
    <title>捷途软件--智能办公</title>
    <link href="${ctx}/css/base.css" rel="stylesheet">
    <link href="${ctx}/css/login.css" rel="stylesheet">
    <link href="${ctx}/resources/bootstrap/css/bootstrap.min.css" rel="stylesheet">
    <script type="text/javascript" src="${ctx }/resources/jquery/jquery-1.11.0.min.js"></script>
    <script type="text/javascript" src="${ctx }/resources/jquery/jquery-migrate-1.2.1.min.js"> </script>
    <script type="text/javascript" src="${ctx }/resources/bootstrap/js/bootstrap.min.js"></script>
    <script type="text/javascript" src="${ctx}/resources/easyUI/jquery.easyui.min.js"></script>
    <script type="text/javascript" src="${ctx}/resources/easyUI/easyui-lang-zh_CN.js"></script>
    <link rel="stylesheet" href="${ctx}/resources/easyUI/easyui.css">
    <script type="text/javascript">
        $(function(){
            // 等文档加载完成以后再执行本脚本
            // 给验证码绑定单击事件
            // vimg
            $("#vimg").click(function(){
                $(this).attr("src","${ctx}/createCode?timer="+new Date().getTime());
            }).mouseover(function(){
                $(this).css("cursor","pointer");
            });

            /** 回车键事件
              event :事件源，代表按下的那个按键
            */
            $(document).keydown(function(event){
                if(event.keyCode == 13){
                    $("#login_id").trigger("click");
                }
            });

            /** 1.异步登录功能  */
            $("#login_id").bind("click",function(){
                var userId = $("#userId").val();
                var passWord = $("#passWord").val();
                var vcode = $("#vcode").val();

                // 定义一个校验结果
                var msg = "";
                if(!/^\w{2,20}$/.test(userId.trim())){
```

```javascript
                    msg = "登录名必须是2~20个字符";
                }else if(!/^\w{6,20}$/.test(passWord)){
                    msg = "密码必须是6~20个字符";
                }else if(!/^\w{4}$/.test(vcode)){
                    msg = "验证码格式不正确";
                }

                if(msg!=""){
                    // 校验失败
                    $.messager.alert("登录提示","<span style='color:red;'>"+msg+"</span>","error");
                    return ; // 结束程序
                }

                var params = $("#loginForm").serialize();

                /** 发起异步请求登录 */
                $.ajax({
                    url:"${ctx}/loginAjax",
                    type: "post",
                    dataType : "json",
                    data : params ,
                    async : true ,   // 是异步还是异步中的同步
                    success : function(data){
                        if(data.status == 1){
                            /** 跳转到主界面 */
                            window.location = "${ctx}/oa/main";
                        }else{
                            $("#vimg").trigger("click");
                            $.messager.alert("登录提示","<span style='color:red;'>"+data.tip+"</span>","error");
                        }
                    },error : function(){
                        $.messager.alert("登录提示","<span style='color:red;'>您登录失败了</span>","error");
                    }
                })

            })

    </script>
</head>
<body>
    <div class="login-hd">
        <div class="left-bg"></div>
        <div class="right-bg"></div>
        <div class="hd-inner">
            <span class="logo"></span>
            <span class="split"></span>
            <span class="sys-name">智能办公平台</span>
        </div>
    </div>
    <div class="login-bd">
        <div class="bd-inner">
            <div class="inner-wrap">
                <div class="lg-zone">
                    <div class="lg-box">
                        <div class="panel-heading" style="background-color: #11a9e2;">
                            <h3 class="panel-title" style="color: #FFFFFF;font-style:
```

```html
italic;">用户登录</h3>
                            </div>
                            <form id="loginForm">
                                <div class="form-horizontal" style="padding-top: 20px;
padding-bottom: 30px; padding-left: 20px;">
                                    <div class="form-group" style="padding: 5px;">
                                        <div class="col-md-11">
                                            <input class="form-control" id="userId" name=
"userId" type="text" placeholder="账号/邮箱">
                                        </div>
                                    </div>
                                    <div class="form-group" style="padding: 5px;">
                                        <div class="col-md-11">
                                            <input class="form-control" id="passWord" name=
"passWord" type="password" placeholder="请输入密码">
                                        </div>
                                    </div>
                                    <div class="form-group" style="padding: 5px;">
                                        <div class="col-md-11">
                                            <div class="input-group">
                                                <input class="form-control " id="vcode" name=
"vcode" type="text" placeholder="验证码">
                                                <span class="input-group-addon" id="basic-addon2">
<img class="check-code" id="vimg" alt="" src="${ctx}/createCode"></span>
                                            </div>
                                        </div>
                                    </div>
                                </div>
                                <div class="tips clearfix">
                                    <label><input type="checkbox" checked="checked">
记住用户名</label>
                                    <a href="javascript:;" class="register">忘记
密码?</a>
                                </div>
                                <div class="enter">
                                    <a href="javascript:;" id="login_id" class="purchaser">
登录</a>
                                    <a href="javascript:;" class="supplier" onClick=
"javascript:window.location='main.html'">重 置</a>
                                </div>
                            </form>
                        </div>
                        <div class="lg-poster"></div>
                    </div>
                </div>
            </div>
            <div class="login-ft">
                <div class="ft-inner">
                    <div class="about-us">
                        <a href="javascript:;">关于我们</a>
                        <a href="javascript:;">法律声明</a>
                        <a href="javascript:;">服务条款</a>
                        <a href="javascript:;">联系方式</a>
                    </div>
                    <div class="address">
地址：广州市天河区车陂大岗路 X 号，XX 大厦 X011
 邮编：510000  
Copyright © 2015 - 2016 疯狂软件-分享知识,传递希
```

```html
望 版权所有</div>
            <div class="other-info">
            建议使用火狐、谷歌浏览器，不建议使用IE浏览器！</div>
        </div>
    </div>
</body>
</html>
<script type="text/javascript">

</script>
```

程序清单：codes/07/oa/src/main/java/org/fkit/oa/identity/controller/LoginController

```java
import java.util.HashMap;
import java.util.Map;
import javax.annotation.Resource;
import javax.servlet.http.HttpSession;
import org.fkit.oa.identity.service.IdentityService;
import org.springframework.stereotype.Controller;
import org.springframework.web.bind.annotation.RequestMapping;
import org.springframework.web.bind.annotation.RequestParam;
import org.springframework.web.bind.annotation.ResponseBody;

@Controller
public class LoginController {

    /** 1.定义业务层对象 */
    @Resource  // by type
    private IdentityService identityService;

    @ResponseBody   // 异步请求的响应结果
    @RequestMapping(value="/loginAjax",produces="application/json; charset=UTF-8")
    public Map<String, Object> login(@RequestParam("userId")String userId,
            @RequestParam("passWord")String passWord
            ,@RequestParam("vcode")String vcode
            ,HttpSession session){
        try {
            Map<String, Object> params = new HashMap<>();
            params.put("userId", userId);
            params.put("passWord", passWord);
            params.put("vcode", vcode);
            params.put("session", session);
            // 获得数据
            Map<String, Object> result = identityService.login(params);
            // 写回客户端
            return result;
        } catch (Exception e) {
            e.printStackTrace();
        }
        return null;
    }
}
```

LoginController 处理登录请求，当用户登录成功后，转入"/oa/main"；用户登录失败，弹出提示信息。

程序清单：codes/07/oa/src/main/java/org/fkit/oa/identity/controller/RequestController

```java
import java.util.List;
import javax.annotation.Resource;
import org.fkit.oa.identity.dto.UserModule;
import org.fkit.oa.identity.service.IdentityService;
import org.springframework.stereotype.Controller;
import org.springframework.ui.Model;
import org.springframework.web.bind.annotation.RequestMapping;
```

```java
@Controller
@RequestMapping("/oa")
public class RequestController {

    /** 1.定义业务层对象 */
    @Resource // by type
    private IdentityService identityService;

    @RequestMapping(value="/login")
    public String requestLogin(){
        System.out.println("登录成功了！");
        return "login";
    }

    @RequestMapping(value="/main")
    public String requestMain(Model model){
        try {
            //查询当前用户拥有的所有模块权限
            List<UserModule> userModules = identityService.getUserPopedomModules();
            model.addAttribute("userPopedomModules", userModules);

        } catch (Exception e) {
            e.printStackTrace();
        }
        return "main";
    }

    @RequestMapping(value="/home")
    public String requestHome(){
        return "home";
    }
}
```

RequestController 中 requestMain 方法响应 "main" 请求，该方法查询当前用户拥有的所有模块权限。

程序清单：codes/07/oa/webapp/WEB-INF/jsp/main.jsp

```jsp
<%@ page language="java" contentType="text/html; charset=UTF-8"
    pageEncoding="UTF-8"%>
<!DOCTYPE html>
<head>
    <meta name="viewport" content="width=device-width, initial-scale=1">
    <title>捷途软件--智能办公</title>
<%@ include file="/WEB-INF/taglib.jsp"%>
<link href="${ctx}/css/base.css" rel="stylesheet">
<link href="${ctx}/css/platform.css" rel="stylesheet">
<link rel="stylesheet" href="${ctx}/resources/easyUI/easyui.css">
<script type="text/javascript" src="${ctx }/resources/jquery/jquery-1.11.0.min.js"></script>
<script type="text/javascript" src="${ctx }/resources/jquery/jquery-migrate-1.2.1.min.js"></script>
<script type="text/javascript" src="${ctx}/resources/easyUI/jquery.easyui.min.js"></script>
<!-- <script type="text/javascript" src="js/menu.js"></script> -->
<script type="text/javascript" src="${ctx}/resources/main.js"></script>
<script type="text/javascript">
    $(function(){
        $('#tt').tabs({
                tabHeight: 40,
                onSelect:function(title,index){
                    var currentTab = $('#tt').tabs("getSelected");
                    if(currentTab.find("iframe") && currentTab.find("iframe").size()){
```

```js
                            currentTab.find("iframe").attr("src",currentTab.find
("iframe").attr("src"));
                        }
                    }
            });
        })

            // 写一个方法, 向 easyUI 中添加面板
            function addPanel(id,url,name){
                name = name.replace(/-/g,"");
                // 判断之前是否已经存在该面板, 如果存在, 则不创建新的面板
                var exist = $('#tt').tabs('exists',name);
                if(exist){
                    // 如果已经存在, 就将该面板选中   exist
                    $('#tt').tabs('select',name);
                    var currentTab = $('#tt').tabs('getTab',name);
                    // 刷新一下界面
                    if(currentTab.find("iframe") && currentTab.find("iframe").size()){
                        currentTab.find("iframe").attr("src",currentTab.find
("iframe").attr("src"));
                    }
                }else{
                    $('#tt').tabs('add',{
                        id:id,
                        title: name,
                        content: '<div style="width:100%;height:100%;"><iframe class="page-iframe" src="${ctx}'+url+'" frameborder="no"  border="no" height="100%" width="100%" scrolling="auto"></iframe></div>',
                        closable: true
                    });
                }
            }

        $(window).resize(function(){
            $('.tabs-panels').height($("#pf-page").height()-46);
            $('.panel-body').height($("#pf-page").height()-76)
        }).resize();

        var page = 0,
            pages = ($('.pf-nav').height() / 70) - 1;

        if(pages === 0){
          $('.pf-nav-prev,.pf-nav-next').hide();
        }
        $(document).on('click', '.pf-nav-prev,.pf-nav-next', function(){

          if($(this).hasClass('disabled')) return;
          if($(this).hasClass('pf-nav-next')){
            page++;
            $('.pf-nav').stop().animate({'margin-top': -70*page}, 200);
            if(page == pages){
              $(this).addClass('disabled');
              $('.pf-nav-prev').removeClass('disabled');
            }else{
              $('.pf-nav-prev').removeClass('disabled');
            }
          }else{
            page--;
            $('.pf-nav').stop().animate({'margin-top': -70*page}, 200);
            if(page == 0){
              $(this).addClass('disabled');
              $('.pf-nav-next').removeClass('disabled');
            }else{
              $('.pf-nav-next').removeClass('disabled');
```

```html
                    }
                }
            })
            function exit(){
                window.location = "${ctx}/logout";
            }
        </script>
    </head>
    <body>
        <div class="container">
            <div id="pf-hd">
                <div class="pf-logo"">
                    <img src="${ctx}/images/main/main_logo.png" alt="logo">
                </div>

                <div class="pf-nav-wrap">
                    <div class="pf-nav-ww">
                        <ul class="pf-nav">
                            <li class="pf-nav-item home" data-menu="sys-manage">
                                <a href="javascript:;">
                                    <span class="iconfont">&#xe63f;</span>
                                    <span class="pf-nav-title">系统管理</span>
                                </a>
                            </li>
                        </ul>
                    </div>
                    <a href="javascript:;" class="pf-nav-prev disabled iconfont">&#xe606;</a>
                    <a href="javascript:;" class="pf-nav-next iconfont">&#xe607;</a>
                </div>


                <div class="pf-user">
                    <div class="pf-user-photo">
                        <img style="width: 40px;height: 40px;" src="${ctx}/images/main/xlei.jpg" alt="">
                    </div>
                    <h4 class="pf-user-name ellipsis">徐磊</h4>
                    <i class="iconfont xiala">&#xe607;</i>

                    <div class="pf-user-panel">
                        <ul class="pf-user-opt">
                            <li>
                                <a href="javascript:;">
                                    <i class="iconfont">&#xe60d;</i>
                                    <span class="pf-opt-name">用户信息</span>
                                </a>
                            </li>
                            <li id="exit">
                                <a href="javascript:exit();">
                                    <i class="iconfont">&#xe60e;</i>
                                    <span class="pf-opt-name">退出</span>
                                </a>
                            </li>
                        </ul>
                    </div>
                </div>
```

```html
            </div>
            <div id="pf-bd">
                <div id="pf-sider">
                    <h2 class="pf-model-name">
                        <span class="iconfont">&#xe64a;</span>
                        <span class="pf-name">信息系统</span>
                        <span class="toggle-icon"></span>
                    </h2>
                    <!-- 展示系统左侧权限树 -->
                    <ul class="sider-nav" id="sider-nav">
                                    <c:forEach items="${userPopedomModules}" var="userModule">
                                       <li>
                                          <a href="javascript:;" >
                                          <span class="iconfont sider-nav-icon">&#xe611;</span>
                                          <span class="sider-nav-title">${userModule.firstModule.name}</span>
                                          <i class="iconfont">&#xe642;</i>
                                         </a>
                                         <ul class="sider-nav-s">
                                            <c:forEach items="${userModule.secondModules }" var="secondModule">
                                                <li><a href="javascript:addPanel('${secondModule.code }' ,'${secondModule.url }','${secondModule.name}');">${secondModule.name}</a></li>
                                            </c:forEach>
                                         </ul>
                                      </li>
                                   </c:forEach>

                   </ul>
                </div>

                <!-- 面板 -->
                <div id="pf-page">
                    <div class="easyui-tabs" id="tt" style="width:100%;height:100%;">
                        <div title="当前用户" id="user" style="padding:10px 5px 5px 10px;">
                            <iframe class="page-iframe" src="${ctx}/oa/home" frameborder="no"  border="no" height="100%" width="100%" scrolling="auto"></iframe>
                        </div>
                    </div>
                </div>
            </div>

            <div id="pf-ft">
                <div class="system-name">
                 <i class="iconfont">&#xe6fe;</i>
                 <span>智能办公平台 v1.0</span>
                </div>
                <div class="copyright-name">
                    <span>Copyright © 2015 - 2016 疯狂软件-分享知识，传递希望 版权所有</span>
                    <i class="iconfont" >&#xe6ff;</i>
                </div>
            </div>
        </div>
    </body>
</html>
```

再次运行 App 类的 main 方法。Spring Boot 项目启动后，在浏览器中输入 URL 测试应用。

```
http://localhost:8080/oa/login
```

登录页面如图 7.2 所示。

图 7.2 登录页面

输入正确的登录名：admin，密码：123456，再输入自动生成的验证码，单击"登录"按钮。登录成功，跳转到如图 7.1 所示的"信息管理系统"页面。

"信息管理系统"主页面分为 3 部分。上面是"信息管理系统"的 logo 和登录的用户名和头像；左边是菜单栏，显示系统的模块；右边主显示框显示登录的用户信息。菜单数据和用户信息都来自于数据库的原始数据。

单击左侧菜单"系统管理"，展开"系统管理"菜单，菜单下面有 3 个大的模块："用户管理"、"菜单管理"和"角色管理"。

菜单管理可以添加和删除模块；角色管理可以添加、删除、修改角色，绑定操作模块并分配给用户；用户管理可以添加、删除、修改、查询、禁用和激活用户。这一切都可以通过角色来进行权限分配和管理。

7.6.3 菜单管理

单击左侧菜单"系统管理"下面的"菜单管理"，主显示框跳转到"菜单管理"界面，如图 7.3 所示。

页面中展现"系统模块树"结构。分别点开"用户管理"、"菜单管理"和"角色管理"，系统初始化定义的管理模块如图 7.4 所示。

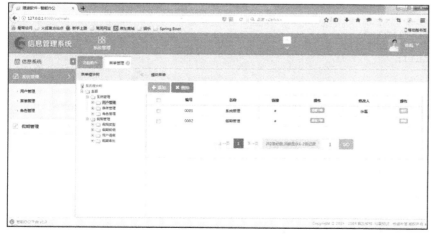

图 7.3 菜单管理页面

图 7.4 菜单模块树

所有的菜单模块树中的模块都来自"oa_id_module"表中的原始数据。

> **注意**
> 用户也可以使用系统的添加菜单功能，增加自定义模块功能，扩展功能，这是这个"信息管理系统"最具商业价值的地方。"假期管理"模块就是留给读者去扩展的功能模块。

ModuleController 是处理"菜单管理"功能的控制器。

程序清单：codes/07/oa/src/main/java/org/fkit/oa/identity/controller/ModuleController

```java
import java.util.List;
import javax.annotation.Resource;
import org.fkit.common.util.pager.PageModel;
import org.fkit.oa.identity.domain.Module;
import org.fkit.oa.identity.service.IdentityService;
import org.fkit.oa.identity.vo.TreeData;
import org.springframework.stereotype.Controller;
import org.springframework.ui.Model;
import org.springframework.web.bind.annotation.RequestMapping;
import org.springframework.web.bind.annotation.ResponseBody;

/**
 * 模块菜单
 */
@Controller
@RequestMapping("/identity/module")
public class ModuleController {

    @Resource // by type
    private IdentityService identityService;

    @RequestMapping(value="/mgrModule")
    public String mgrModule(){
        return "identity/module/mgrModule";
    }

    @ResponseBody
    @RequestMapping(value="/loadAllModuleTrees",produces="application/json;charset=UTF-8")
    public List<TreeData> loadAllModuleTrees(){
        try {
            return identityService.loadAllModuleTrees();
        } catch (Exception e) {
            e.printStackTrace();
            return null;
        }
    }
    @RequestMapping(value="/getModulesByParent")
    public String getModulesByParent(String parentCode,PageModel pageModel,Model model){
        try {
            List<Module> sonModules = identityService.getModulesByParent(parentCode,pageModel);
            model.addAttribute("modules", sonModules);
            model.addAttribute("parentCode", parentCode);
        } catch (Exception e) {
            e.printStackTrace();
        }
        return "identity/module/module";
    }
```

```java
@RequestMapping(value="/deleteModules")
public String deleteModules(String ids,Model model){
    try {
        identityService.deleteModules(ids);
        model.addAttribute("tip", "删除成功");
    } catch (Exception e) {
        model.addAttribute("tip", "删除失败");
        e.printStackTrace();
    }
    return "forward:/identity/module/getModulesByParent";
}
@RequestMapping(value="/addModule")
public String addModule(String parentCode,Module module ,Model model){
    try {
        identityService.addModule(parentCode,module);
        model.addAttribute("tip", "添加成功");
        model.addAttribute("parentCode", parentCode);
    } catch (Exception e) {
        model.addAttribute("tip", "添加失败");
        e.printStackTrace();
    }
    return "identity/module/addModule";
}
@RequestMapping(value="/updateModule")
public String updateModule(Module module ,Model model){
    try {
        identityService.updateModule(module);
        model.addAttribute("tip", "修改成功");
    } catch (Exception e) {
        model.addAttribute("tip", "修改失败");
        e.printStackTrace();
    }
    return "identity/module/updateModule";
}
@RequestMapping(value="/showAddModule")
public String showAddModule(String parentCode,Model model){
    model.addAttribute("parentCode", parentCode);
    return "identity/module/addModule";
}
@RequestMapping(value="/showUpdateModule")
public String showUpdateModule(Module module ,Model model){
    try {
        module = identityService.getModuleByCode(module.getCode());
        model.addAttribute("module", module);
    } catch (Exception e) {
        e.printStackTrace();
    }
    return "identity/module/updateModule";
}
}
```

/webapp/WEB-INF/jsp/identity/module 目录下是和"菜单管理"相关的所有 JSP 页面。因为涉及的 JSP 页面代码太多，限于篇幅，此处不一一列举。

7.6.4 角色管理

单击左侧菜单"系统管理"下的"角色管理"，主显示框跳转到"角色管理"界面，如图 7.5 所示。

图 7.5　角色管理页面

RoleController 是处理"角色管理"功能的控制器。

程序清单：codes/07/oa/src/main/java/org/fkit/oa/identity/controller/RoleController

```java
import java.util.List;
import javax.annotation.Resource;
import org.fkit.common.util.pager.PageModel;
import org.fkit.oa.identity.domain.Role;
import org.fkit.oa.identity.domain.User;
import org.fkit.oa.identity.service.IdentityService;
import org.springframework.stereotype.Controller;
import org.springframework.ui.Model;
import org.springframework.web.bind.annotation.RequestMapping;

@Controller
@RequestMapping("/identity/role")
public class RoleController {
    /** 1.定义业务层对象 */
    @Resource // by type
    private IdentityService identityService;

    @RequestMapping(value="/selectRole")
    public String selectRole(PageModel pageModel,Model model){
        try {
            List<Role> roles = identityService.getRoleByPager(pageModel);
            model.addAttribute("roles", roles);
        } catch (Exception e) {
            e.printStackTrace();
        }
        return "identity/role/role";
    }

    @RequestMapping(value="/addRole")
    public String addRole(Role role,Model model){
        try {
            identityService.addRole(role);
            model.addAttribute("tip","添加成功");
        } catch (Exception e) {
            model.addAttribute("tip","添加失败");
            e.printStackTrace();
        }
        return "identity/role/addRole";
    }
```

```java
@RequestMapping(value="/deleteRole")
public String deleteRole(String ids,Model model){
    try {
        identityService.deleteRole(ids);
        model.addAttribute("tip","删除成功");
    } catch (Exception e) {
        model.addAttribute("tip","删除失败");
        e.printStackTrace();
    }
    return "forward:/identity/role/selectRole";
}

@RequestMapping(value="/showAddRole")
public String showAddRole(){
    return "identity/role/addRole";
}

@RequestMapping(value="/showUpdateRole")
public String showUpdateRole(Role role , Model model){
    try {
        role = identityService.getRoleById(role.getId());
        model.addAttribute("role", role);
    } catch (Exception e) {
        e.printStackTrace();
    }
    return "identity/role/updateRole";
}

@RequestMapping(value="/updateRole")
public String updateRole(Role role , Model model){
    try {
        identityService.updateRole(role);
        model.addAttribute("tip", "修改成功");
    } catch (Exception e) {
        model.addAttribute("tip", "修改失败");
        e.printStackTrace();
    }
    return "identity/role/updateRole";
}

@RequestMapping(value="/selectRoleUser")
public String selectRoleUser(Role role ,PageModel pageModel, Model model){
    try {
        /** 查询属于这个角色的所有用户 */
        List<User> users = identityService.selectRoleUser(role,pageModel);
        role = identityService.getRoleById(role.getId());
        model.addAttribute("users", users);
        model.addAttribute("role", role);
    } catch (Exception e) {
        e.printStackTrace();
    }
    return "identity/role/bindUser/roleUser";
}

@RequestMapping(value="/showBindUser")
public String selectNotRoleUser(Role role ,PageModel pageModel, Model model){
    try {
        /** 查询不属于这个角色的所有用户 */
        List<User> users = identityService.selectNotRoleUser(role,pageModel);
        role = identityService.getRoleById(role.getId());
```

```
            model.addAttribute("users", users);
            model.addAttribute("role", role);
        } catch (Exception e) {
            e.printStackTrace();
        }
        return "identity/role/bindUser/bindUser";
    }


    @RequestMapping(value="/bindUser")
    public String bindUser(Role role,String ids, Model model){
        try {
            identityService.bindUser(role,ids);
            model.addAttribute("tip","绑定成功");
        } catch (Exception e) {
            model.addAttribute("tip","绑定失败");
            e.printStackTrace();
        }
        return "forward:/identity/role/showBindUser";
    }


    @RequestMapping(value="/unBindUser")
    public String unBindUser(Role role,String ids, Model model){
        try {
            identityService.unBindUser(role,ids);
            model.addAttribute("tip","解绑成功");
        } catch (Exception e) {
            model.addAttribute("tip","解绑失败");
            e.printStackTrace();
        }
        return "forward:/identity/role/selectRoleUser";
    }
}
```

/webapp/WEB-INF/jsp/identity/role 目录下是和"角色管理"相关的所有 JSP 页面。因为涉及的 JSP 页面代码太多，限于篇幅，此处不一一列举。

单击左侧菜单"系统管理"下的"用户管理"，主显示框跳转到"用户管理"界面，如图 7.6 所示。

图 7.6 用户管理页面

用户当前只可进行查询操作和预览操作，可以给用户赋予增加操作、删除操作、修改操作和激活用户操作，这就需要使用角色权限管理功能。

单击左侧菜单"系统管理"下的"角色管理"，主显示框跳转到"角色管理"界面，如图7.5所示。找到角色名称为"系统管理员"的一列数据，单击"绑定操作"，跳转到"菜单模块树"，单击"系统管理"下的"用户管理"，进入"模块菜单"赋予权限，如图7.7所示。

图 7.7　给角色授予权限

现在是为名称为"系统管理员"的角色绑定可进行的操作，选中"用户管理"下的所有操作，单击"绑定权限"按钮，系统提示"绑定成功"。则"系统管理员"角色可以操作用户的增加、删除、修改、查询、预览和激活用户操作。打开 MySQL 数据库，观察 oa_id_popedom 表，会发现新增了 6 条数据，表中的 role_id 表示关联 oa_id_role 表的权限编号，module_code 表示关联 oa_id_module 表的模块编号，operta_code 表示用户可操作的模块编号。

由于测试原始数据，登录的"admin"用户就是"系统管理员"，所以此处不用再做任何操作。退出系统，再次使用登录名：admin，密码：123456，登录系统。单击"用户管理"，进入"用户管理列表"页面，可以看到，当前页面已经多了"添加"、"删除"、"修改"按钮，"激活"开关按钮也是可用的，说明用户权限分配成功，如图7.8所示。

图 7.8　用户赋予权限成功

7.6.5 用户管理

UserController 是处理 "用户管理" 功能的控制器。

程序清单：codes/07/oa/src/main/java/org/fkit/oa/identity/controller/UserController

```java
import java.util.List;
import javax.annotation.Resource;
import javax.servlet.http.HttpServletRequest;
import javax.servlet.http.HttpSession;
import org.fkit.common.util.pager.PageModel;
import org.fkit.oa.identity.domain.User;
import org.fkit.oa.identity.service.IdentityService;
import org.springframework.stereotype.Controller;
import org.springframework.ui.Model;
import org.springframework.util.StringUtils;
import org.springframework.web.bind.annotation.RequestMapping;
import org.springframework.web.bind.annotation.ResponseBody;

@Controller
@RequestMapping("/identity/user")
public class UserController {
    /** 1.定义业务层对象 */
    @Resource // by type
    private IdentityService identityService;

    @RequestMapping(value="/updateSelf")
    public String updateSelf(User user,Model model,HttpSession session){
        try {
            identityService.updateSelf(user,session);
//            session.setAttribute(OaContants.USER_SESSION, identityService.getUserById(user.getUserId()));
            model.addAttribute("tip", "修改成功！");
        } catch (Exception e) {
            model.addAttribute("tip", "修改失败！");
            e.printStackTrace();
        }
        return "home";
    }

    @RequestMapping(value="/selectUser")
    public String selectUser(User user ,HttpServletRequest request, PageModel pageModel,Model model){
        try {
            /** 1.自己处理 ：只需处理 get 请求的参数
             *    post 请求的参数不会乱码
             */
            if(request.getMethod().toLowerCase().indexOf("get")!=-1){
                if(user!=null && !StringUtils.isEmpty(user.getName())){
                    String name = user.getName();
                    /**
                     * 如果不用 UTF-8 处理，返回到浏览器的数据会是乱码
                     */
                    name = new String(name.getBytes("ISO-8859-1") , "UTF-8");
                    user.setName(name);
                }
            }

            List<User> users = identityService.getUsersByPager(user,pageModel);
            model.addAttribute("users", users);
        } catch (Exception e) {
            e.printStackTrace();
```

```java
        return "identity/user/user";
    }

    @RequestMapping(value="/deleteUser")
    public String deleteUser(String ids ,Model model){
        try {
            /** 批量删除   */
            identityService.deleteUserByUserIds(ids);
            model.addAttribute("tip", "删除成功");
        } catch (Exception e) {
            e.printStackTrace();
            model.addAttribute("tip", "删除失败");
        }
        return "forward:/identity/user/selectUser";
    }

    @RequestMapping(value="/showAddUser")
    public String showAddUser(){
        return "identity/user/addUser";
    }

    @ResponseBody
    @RequestMapping(value="/isUserValidAjax")
    public String isUserValid(String userId){
        try {
            return identityService.isUserValidAjax(userId);
        } catch (Exception e) {
            e.printStackTrace();
            return null;
        }
    }

    @RequestMapping(value="/addUser")
    public String addUser(User user,Model model){
        try {
            /** 批量删除   */
            identityService.addUser(user);
            model.addAttribute("tip", "添加成功");
        } catch (Exception e) {
            e.printStackTrace();
            model.addAttribute("tip", "添加失败");
        }
        return "identity/user/addUser";
    }

    @RequestMapping(value="/updateUser")
    public String updateUser(User user,Model model){
        try {
            /** 批量删除   */
            identityService.updateUser(user);
            model.addAttribute("tip", "修改成功");
        } catch (Exception e) {
            e.printStackTrace();
            model.addAttribute("tip", "修改失败");
        }
        return "identity/user/updateUser";
    }

    @RequestMapping(value="/activeUser")
    public String activeUser(User user,Model model){
```

```java
        try {
            /** 激活用户 */
            identityService.activeUser(user);
            model.addAttribute("tip", user.getStatus()==1?"激活成功":"冻结成功");
        } catch (Exception e) {
            e.printStackTrace();
            model.addAttribute("tip", user.getStatus()==1?"激活失败":"冻结失败");
        }
        return "forward:/identity/user/selectUser";
    }

    @RequestMapping(value="/showPreUser")
    public String showPreUser(String userId,Model model){
        try {
            /** 批量删除 */
            User user = identityService.getUserById(userId);
            model.addAttribute("user", user);
        } catch (Exception e) {
            e.printStackTrace();
        }
        return "identity/user/preUser";
    }

    @RequestMapping(value="/showUpdateUser")
    public String showUpdateUser(String userId,Model model){
        try {
            /** 批量删除 */
            User user = identityService.getUserById(userId);
            model.addAttribute("user", user);
        } catch (Exception e) {
            e.printStackTrace();
        }
        return "identity/user/updateUser";
    }
}
```

在用户信息列表页面可以通过姓名、手机号码、部门、职位进行用户查询，该查询功能支持模糊查询。分别输入要查询的信息，单击"查询"按钮，就可以进行查询，如图7.9所示。

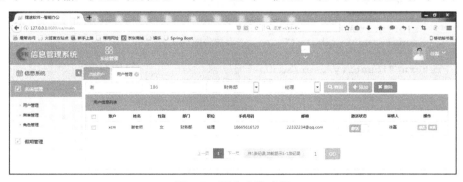

图7.9 用户查询功能

单击"添加"按钮，弹出"添加用户"面板，如图7.10所示。

填写用户信息，单击"添加"按钮，系统提示"添加成功"。单击右上角的"×"，返回"用户信息列表"，可以看到刚刚添加的用户信息，但是用户激活状态为"冻结"。

图 7.10 添加用户

选择新增用户记录最后一列"操作"中的"添加"按钮,可以进行用户修改操作,如图 7.11 所示。

图 7.11 修改用户

选中新增用户记录,单击"删除"按钮,弹出删除用户提示框,可以进行用户删除操作,如图 7.12 所示。

图 7.12 删除用户

单击新增用户激活状态一列的开关按钮,可以激活用户,如图7.13所示。

图7.13 激活用户

接下来给新增用户授权。单击左侧菜单"系统管理"下的"角色管理",主显示框跳转到"角色管理"页面,如图7.5所示。

如果想将"系统管理员"角色分配给新增用户,单击"绑定用户",跳转到"用户授权"页面,如图7.14所示。

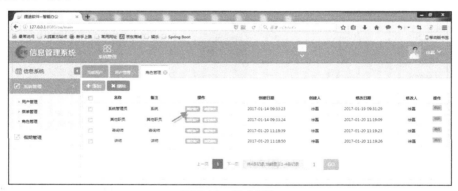

图7.14 用户授权

单击"绑定用户",跳转到"系统管理员"页面,如图7.15所示。

图7.15 "系统管理员"页面

"系统管理员"页面只有一条记录,说明现在系统中只有一个用户拥有"系统管理员"角色权限。单击"绑定用户"按钮,选择要绑定的用户,如图 7.16 所示。

图 7.16 绑定用户

单击"绑定"按钮,系统提示"绑定成功"。单击右上角的"×",返回"系统管理员"页面,可以看到新增用户成功绑定"系统管理员"角色,如图 7.17 所示。

图 7.17 成功绑定系统管理员

退出系统,再次使用登录名:xiaowenji,密码:123456,登录系统,新增用户也拥有"系统管理员"的所有权限。

▶▶ 7.6.6 功能扩展

"信息管理系统"最大的特点是可以无限扩展功能。例如现在公司需要增加一个"财务管理"功能。只需增加菜单,分配权限,再进行"财务管理"功能的开发即可。

单击左侧菜单"系统管理"下的"菜单管理",主显示框跳转到"菜单管理"界面,如图 7.3 所示。单击"添加"按钮,添加"财务管理"菜单,如图 7.18 所示。

> **注意**
> 操作地址表示功能模块的 URL 链接地址,对应数据库 oa_id_module 表的 url 列。财务管理模块是顶级菜单,所以不需要给出具体链接地址,使用"#"表示。

图 7.18　添加顶级菜单

单击"添加"按钮，菜单添加成功，如图 7.19 所示。

图 7.19　添加菜单成功

单击"菜单模块树"中的"财务管理"菜单，显示"财务管理"菜单下没有子菜单。单击"添加"按钮，添加一个上层菜单"工资管理"，如图 7.20 所示。

图 7.20　添加上层菜单

在弹出的"添加模块"面板中添加一个上层菜单"工资管理"，如图 7.21 所示。

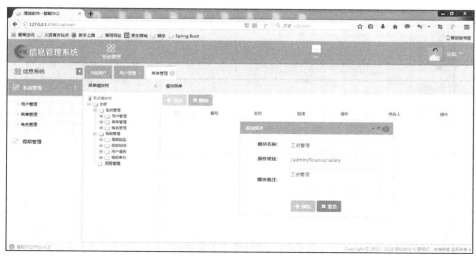

图 7.21　上层菜单"工资管理"

单击"添加"按钮，添加"工资管理"菜单成功，如图 7.22 所示。

图 7.22　上层菜单添加成功

单击"财务管理"下的"工资管理"菜单，显示"工资管理"菜单下没有子菜单。单击"添加"按钮，添加子菜单"工资查询"，如图 7.23 所示。

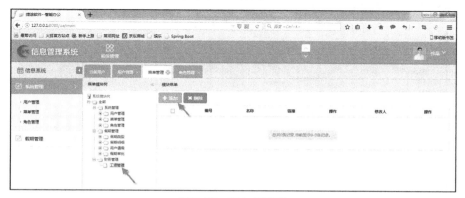

图 7.23　添加子菜单

在弹出的"添加模块"面板中添加一个子菜单"工资查询"，如图 7.24 所示。

图 7.24　子菜单"工资查询"

单击"添加"按钮，添加"工资查询"子菜单成功，如图 7.25 所示。

图 7.25　添加子菜单成功

菜单添加完成后，需要进行用户绑定操作，让用户能够操作新添加的菜单。选择"系统管理"下的"角色管理"，找到角色名称为"系统管理员"的一列数据，单击"绑定操作"，跳转到"菜单模块树"，如图 7.26 所示。

图 7.26　绑定新菜单操作

单击"工资管理"菜单，会出现可以绑定的操作，此处只有一个"工资查询"。单击"绑定权限"按钮，系统提示绑定权限成功，如图 7.27 所示。

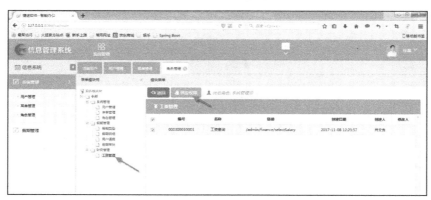

图 7.27 绑定工资管理菜单操作

退出系统，再次使用"系统管理员"用户登录，可以看到左边的菜单栏多了"财务管理"菜单，点开"财务管理"菜单，可以看到子菜单"工资管理"，如图 7.28 所示。

图 7.28 菜单绑定成功

新增"财务管理"功能菜单已经完成，当然，"工资管理"模块里面的"工资查询"等功能的业务逻辑和 JSP 页面还需要读者自行开发。如果有兴趣，还可以继续增加例如"报销管理"、"报表管理"等功能，无限扩展"信息管理系统"。

7.7 本章小结

本章详细介绍了一个完整的 Spring Boot 项目：信息管理系统，在此基础上可以扩展出企业的 IMS 系统、OA 系统等。因为企业平台本身的复杂性，所以本项目涉及的表达到 6 个，而且各个模块的业务逻辑也比较复杂，这些对初学者可能有一定难度，但只要读者先认真阅读本书前面章节所介绍的知识，并结合本章的讲解，再配合配套代码中的案例，一定可以掌握本章所介绍的内容。

本章所介绍的 Spring Boot 应用综合了前 6 章介绍的所有知识，因此本章内容既是对前面知识点的回顾和复习，也是将理论知识应用到实际开发的典范。一旦掌握本章案例的开发方法，就会对 Spring Boot 企业应用的开发产生豁然开朗的感觉。